Big-(Wo)men, Tyrants, Chiefs, Dictators, Emperors and Presidents

"Intentio vero nostra est manifestare ea quae sunt sicut sunt"
"In truth, it is our intention to describe the things which are exactly as they are"
Frederick II Hohenstaufen "Stupor Mundi".

Francesco dell'Isola

Big-(Wo)men, Tyrants, Chiefs, Dictators, Emperors and Presidents

Towards the Mathematical Understanding of Social Groups

 Springer

Francesco dell'Isola
Università di Roma La Sapienza
Rome, Italy

Università dell'Aquila
L'Aquila, Italy

ISBN 978-981-13-9481-2   ISBN 978-981-13-9479-9   (eBook)
https://doi.org/10.1007/978-981-13-9479-9

This Springer imprint is published by the registered company Springer Nature Singapore Pte Ltd.
The registered company address is: 152 Beach Road, #21-01/04 Gateway East, Singapore 189721, Singapore

This work is dedicated to the memory of my beloved uncle Luigi De Luca.
He was Professor of Classical Languages—Grammar and Literature and Principal of High Schools and transmitted to many generations of students the pleasure of understanding the logic intrinsic in reality. He was capable to explain to everybody, in a precise and rigorous way, every abstract idea, even the most difficult. He has taught me nearly every concept which I later needed in my scientific career, including the basics of set theory. I will never forget his lecture about Giambattista Vico, where he let me understand Vico's dream of transforming history into the phenomenological evidence predicted by A New Science (Una Scienza Nuova). I hope that his pedagogical spirit will revive in this work.

Foreword

When I met the author, he was graduating in physics and I was the professor lecturing his class in measure theory. Since then, I have never ceased to be surprised by his enthusiasm for studies. This was enough, for me, to have a good reason for reading what he has to say, and indeed the exposition in his book is accurate and pervaded by a contagious enthusiasm.

The manipulations that, as for example in the last years in Italy, the electoral systems underwent for purposes not always aimed at the collective well-being show that democracy is difficult to achieve. The situation in which the world navigates today suggests that democracy is sometimes not the best way to make a decision. The persuasive means that criminal and governmental organisations use to guide people's opinions towards their own interests are so refined and achieve such effects as to suggest that, in reality, a dictator actually does exist very often. Sometimes, when faced with the inability of a democratic decision, a dictator is even desired. Therefore, an analysis such as that presented by Francesco dell'Isola, which is easy to read but is facing the problems with competence, depth, rigour and a pinch of imagination, is to be welcomed.

In fact, his analysis is difficult to find elsewhere. It contains several historical references, widely developed and consistent with the subject of discussion, which are conducted in such a way that the reader, while reading this book, builds and refines his knowledge and his understanding of the subject. Thus, among curious, sometimes-questionable!-but-never-trivial speculations, stories and observations, one arrives at more than half of the volume, before finding a parenthesis on the mathematical theory of games, which at this point seems very opportune. Nothing highly technical, like almost the whole essay (except, perhaps, the second appendix on Arrow's theorem).

The text deserves to be read. It could leave the reader not completely satisfied, as the subject is difficult and the limit on the number of pages was stringent, but it will increase his intellectual curiosity towards some challenging parts of modern science. That was not a small feat.

Vincenzo Aversa
Emeritus Professor of Mathematics
Applied to Social and Economical Sciences
Università di Napoli Federico II
Naples, Italy

Preface

Exact Sciences, as evolved from Hellenistic Science, are the tool, which humans have invented to understand all natural phenomena.

Exact sciences are based on the formulation of mathematical models, which "mimic" natural systems. Using mathematical techniques, one then gets the solution of some "problems" which, finally, allow for the prediction and control of natural phenomena.

In this book, we try to present, in a friendly way, the mathematical ideas which made us understand some important aspects of the dynamics of social groups. We believe that they are very similar to those used to describe the behaviour of Lagrangian dynamical systems! This similarity may attract the attention of the layman having the curiosity to understand the intrinsic unity of natural phenomena.

The reader is warned: although the scientific method has been extremely successful in the description of physical phenomena, human beings often refrain from applying it to the study of themselves as species, social groups or individuals. However, also human behaviour is based on specific natural laws. In this work, we dare to overcome the just mentioned taboo by exploiting the visionary understanding of the structure of social groups gained thanks to the intellectual inventions of two giants in applied mathematics: *Le Marquis de Condorcet* and *Kenneth Arrow*.

Some questions are very important if one wants to understand rationally the structure of human groups:

Why power is always concentrated in the hands of few?
Why every social group has always a chief?
Is it possible to establish rules for a true democracy?
Why men and women incessantly look for power?
Are there differences between the exercise of power by men and women?
What about The Platonic Republic, i.e. a human group directed by Philosophers?
Will it remain a dream forever?

Did humans have always refused to study themselves and their societies?

Not always! Indeed, in the period immediately before French Revolution and during the revolution itself humanity has probably experienced the greatest cultural and scientific advancement ever. The dream of French revolutionaries was to employ rationality in every aspect of intellectual activity, including the study of society and human behaviour.

Marie Jean Antoine Nicolas de Caritat, marquis de Condorcet has been one of the main scientists and mathematicians of that period. He tried to use his outstanding mathematical skills to design laws and societies. Of course, the optimisation of this design, in the intention of Condorcet, was aimed to reach the general happiness and the highest degree of human progress. In 1785, Condorcet wrote the *Essai sur l'application de l'analyse à la probabilité des décisions rendues à la pluralité des voix* (Essay on the Application of Analysis to the Probability of Majority Decisions). This is maybe one of the most important works in the history of applied mathematics and one of the first which applied mathematics to social sciences: in it, one finds one fundamental result and one important conjecture.

Condorcet's main result is called, since then,

The Condorcet's Jury Theorem.

It states that:

> *if every member of a voting group is "more clever than average" (i.e. if he has more chances to cast the "correct" vote than to cast the "wrong" one) then the voting group is "much more clever than the average" higher and higher is the number of its members.*

This result seems to be a strong argument in favour of democracy. The reader should, however, pay careful attention to the hypotheses of the theorem.

It happens very often that applied mathematicians (and their epigones) forget about the hypotheses of a theorem and assume that the theses of proven theorems are always true. Instead, one can only state that: if the theorem's condition is true, then its consequence is assured. This is the structure of an implication. Unfortunately, the certitude of mathematics has been, too often, misunderstood. Mathematics never claims that a consequence is unconditionally true. Instead, it claims that a thesis is always true when the hypotheses of the implication are true.

The important hypothesis in the Jury Theorem is that «every member of a voting group is "more clever than average"». Therefore, the democratic choice is more effective than the choice of every single elector under this condition only. Condorcet himself was aware of the fact that, usually, the electoral bodies do not verify his hypothesis.

Unfortunately, even if one manages (and nobody knows how to do this!) to form an electoral body where all electors are "more clever than the average", usually the decision to be made is not between two options. This plurality of options is one of the hypotheses of the so-called

Condorcet Conjecture.

Its informal statement we quote here:

> *if the alternatives for a social choice are more than two, then it is not possible to establish a rule for obtaining such a social choice from the individuals' choices which is truly democratic.*

In a specific sense, Condorcet was right. Indeed, a theorem, whose formulation makes his conjecture precise, was proven rigorously by Arrow many years later (during XX century).

Before being killed by the Revolution, which he had greatly supported, Condorcet wrote the *Esquisse d'un tableau historique des progrès de l'esprit humain* (Sketch for a Historical Picture of the Progress of the Human Spirit) which is surely one of the major texts of the enlightenment and of historical thought. In this work, Condorcet narrates the history of progress in science, civilisation and technology and argues that there is a strict connection between scientific progress and the acceptance and respect of human rights and justice: he dreams a future and rational society whose structure is shaped by means of the knowledge acquired with science. Condorcet was clearly a follower of Plato's utopia.

The spirit of Condorcet has been fully recovered by Arrow.

They both maintain that a careful and objective classification of observable social phenomena is required, together with a serious mathematical modelling effort, if one wants to understand the reasons for which human societies "are organised in the way in which they are organised".

Hopefully, the reader will agree that, also in this kind of investigations, we must fully accept the spirit of that declaration by Frederick II Hohenstaufen, which we have used as epigraph for this essay.

In order to have some simplified phenomena to study and to model, we will also describe some examples of social groups, which are simpler than human societies. We refer to those groups, which are constituted by chimpanzees, gorillas and bonobos: indeed, they represent a less complex version of human groups which show, however, some of their main features. The phenomena observed and the theories developed by evolutionary ethologists and mathematicians will be described, in this essay, by refraining to present explicitly their most formal mathematical structure or the most technical details. We however believe that it is possible, notwithstanding this choice, to convey the main ideas, which constitute the present knowledge in the field.

The mathematical theory of leadership is far from being complete and exhaustive.

Condorcet's and Arrow's results opened a new field of intellectual challenges to be explored. Many new open problems are confronting now their successors. Moreover, the phenomenology of the leadership is nearly completely unexplored and the rigorous description of its more important aspects is far from being obtained with methods as precise as those used in the mathematical analysis developed by Arrow.

To make clear the difficulties to be faced by the new generations of mathematicians and social scientists, some stories, characters and behaviours will be described in this essay. These characters will be embedded in their social context, in order to try to understand better the reasons of their existence. When it is possible, these stories are interrupted by some explanations based on what it is believed now to be the most reasonable and updated explanation of the described behaviours. Our final intent is to try to spread, also in our *époque*, those ideas which dominated the century of enlightenment, obviously updated in their most advanced forms.

L'esprit des Lumières **(The spirit of Enlightenment) seems to be more and more needed, especially nowadays, to direct human history in the right direction.**

Rome/L'Aquila, Italy   Francesco dell'Isola

Acknowledgements

I did not try to hide my cultural roots, in writing this essay. I was educated in Magna Graecia, where a wonderful melting pot mixed Greek philosophy, Roman pragmatism, Longobard proud sense of freedom, Byzantine duplicity and culture, Arab initiative, inventive and tolerance, Normans loyalty and determination, French sophisticated traditions and sense of State, Spanish opportunism, Piedmontese administrative ideas. My English style reflects my education and evokes the European common Greek and Latin roots. I did not look for sophistication, and I simply tried to maintain my original way of expressing, when the English dictionaries allowed me to do so.

I must also thank all those persons, whose name is not suitable to list here, as one may need too many pages, who allowed me to understand the true meaning of Condorcet Conjecture and Arrow's theorem.

Since I was a child, I was surprised by the apparently inexplicable behaviour of humans. Therefore, in order to understand why they are doing that which they are doing, I was obliged to discover the most beautiful part of modern mathematics.

A scientific study of human behaviour is possible: we must develop this part of knowledge without being afraid of what we could discover and without taboos. Once more Frederick II Hohenstaufen is showing us the right way.

Reviewer Note

Big-(Wo)men, Tyrants, Chiefs, Dictators, Emperors and Presidents.
Towards the mathematical understanding of social groups.
By Francesco dell'Isola

Reviewed by Nicola L. Rizzi, University Roma Tre

Note: Quotations from the book are reported in *italic*; quotations from another author in `courier`.

The author starts with the claim that every social group cannot survive without a dictator (or leader)—the meaning of this word is explained in Appendix 1—and the aim of his book is clearly declared at its beginning, as follows:

> *Why a dictator must always exist?*
> *Arrow's theorem, based on the conjecture of Condorcet, provides us with a precise and rational answer to these questions.*

This statement is discussed, essentially, from the point of view of a mathematician with the help of the Arrow and Nash theorems applied, in a sense, to the Condorcet theory.

Nevertheless, the book is not a technical one because, without missing the mathematical rigour when necessary, it is written in layman's terms and full of examples that help the reader to grasp, without great effort, also complex concepts.

The role and the dynamics of the leadership in many environments and situation, namely zoology, history, academy, family and so on, is investigated.

For each one of the cases, the author gives prototypal examples referred, e.g. for zoology to the primate groups; for history to the Italian process of unity; for academy to some fictional examples of academic behaviour.

It must be stressed that the discussion of the examples is always based on a deep knowledge of the matter being, in many cases, the result of original researches with the evidence of some aspects, generally disregarded, that question the clichés by original and upstream insight. This, in my opinion, is one of the most interesting features of the book.

The reading of the book is a long and exciting travel in which the reader is accompanied by honourable men like Carlo Filangeri, filthy figures like Liborio Romano or archetypical characters like Don Pasquale.

At the end of this travel, pointing out the tendency of the humankind *to enslave themselves,* the author goes further and comes up with the concept of Brittle Morality (or Brittle Honesty) of the humans as the primary cause of the existence of dictators, opposed to the (very small diffuse) resilience and ductility that is the necessary condition for opposing resistance to tyrants.

This is synthesised in the statement

> *As suggested by La Boétie the only way to stop the power of the dictators belonging to the class so beautifully exemplified by Don Pasquale and Caesar Sacristy is to stop having any interaction with them, by refusing any contact and compromise.*

and the author comes to the following pessimistic (but we have to recognise full realistic) conclusion

> *Unfortunately this strategy is not easy to implement, as its success requires the support of a large percentage of group members.*

This conclusion is worsened by the observation that

> *There are, however, many Big-Men and Big-Women who altruistically worked for improving the quality of life of their group. Everybody did meet in his life this kind of person. The most miserable among us did exclusively exploit them and their help.*

At this point, the reader could conclude that the book is the expression of a very pessimistic view of life. Actually, this is not the true conclusion of the book as the author leaves us a hope when, in the final part writes

> *For concluding this essay we want to explicitly state that the search for democracy should not be considered a hopeless endeavor*

in force of the following explanation:

> *How can we solve the problem of designing a democracy, given that a single «democratic» function of choice does not exist?*
>
> *We could say that this problem can probably (and hopefully) be solved as Charles-Louis de Secondat, Baron of La Bréde and Montesquieu had imagined: by means of a system of successive approximations!*

In fact, the author writes

> *We do not want here to claim that the models developed up to now by applied mathe-maticians are sufficient to describe, as effectively as done for physical phenomena, also social and economical phenomena. We simply believe that we are in a transitory phase, which is very similar to the one when Galileo started to understand physical natural phenomena by means of geometry and mathematical theorems.*
>
> *We believe that eventually a scientific understanding of social phenomena will be attained: of course it is very difficult to imagine when this result will be actually attained.*

This crucial and final message can be summarised as follows: (i) the hope (or confidence) in the progress of the scientific knowledge (mathematics, for the author) should lead to a better understanding of the social organisation; (ii) the process of *successive approximations* can be a way to improve the social organisation, and this goal can be reached!

The message, in my opinion, gives more remarkable food for thought to the reader, given also that it strengthens and is strengthened by an analogous feeling, coming from another side of the science, the science of language, expressed in the following quotation:

> It is sobering to realize—as I believe we must—how little we have progressed in our knowledge of human beings and society, or even in formulating clearly the problems that might be seriously studied. But there are, I think, a few footholds that seem fairly firm. I like to believe that the intensive study of one aspect of human psychology—human language—may contribute to a humanistic social science that will serve, as well, as an instrument for social action. It must, needless to say, be stressed that social action cannot await a firmly established theory of human nature and society, nor can the validity of the latter be determined by our hopes and moral judgements. The two—speculation and action—must progress as best they can, looking forward to the day when theoretical inquiry will provide a firm guide to the unending, often grim, but never hopeless struggle for freedom and social justice.

(Noam Chomsky: *Language and Freedom*, Lecture at the University Freedom and the Human Sciences Symposium, Loyola University, Chicago, 8–9 January 1970)

Contents

A Dictator Must Always Exist

The layman often hears talking about the power of the reasoning based on mathematical theories. However, many cannot really believe that this kind of reasoning, so far from their direct experience, can be so important in their lives.

One of the purposes of these pages is to demonstrate how some, although seemingly inexplicable, phenomena can be understood (easily!) by using some basic concepts of set theory: that theory whose basic ideas should be, according to the study programmes of almost all Western countries, known to every high school student.

Many Persons Believe that Mathematics Is Too Abstract a Science to Be Really Useful

Mathematics serves to describe, understand, forecast and control the events (phenomena) that occur in our lives. Many are the phenomena that have been understood and dominated, up to now, thanks to the invention of appropriate mathematical models.

In order to ignite the interest of as many readers as possible, we discuss in this essay some very common phenomena in the life of each of us. We refer to the facts of life concerning both the interaction between individuals in social groups and the leading role that some individuals assume in the groups to which they belong. In fact, each individual of the human species lives and acts in one or more social groups and he is forced to interact incessantly with many other individuals, and therefore, all humans have a wide direct experience of this phenomenology. As proven by the modern primatology (see the works by de Waal cited in the Bibliography), chimpanzees, bonobos and gorillas share with us a great part of this experience.

An Example of Physical Phenomena Far from Human Intuition.

It is not easy to persuade a layman about the importance, for example, of the phenomena of electromagnetism and therefore of the Maxwell equations which describe them. Yet these equations play a very important role in our everyday life. It is, in

© Springer Nature Singapore Pte Ltd. 2019
F. dell'Isola, *Big-(Wo)men, Tyrants, Chiefs, Dictators, Emperors and Presidents*,
https://doi.org/10.1007/978-981-13-9479-9_1

fact, difficult to make explicit the relationship which exists between these equations and the ability to distribute electricity everywhere in our cities or to communicate through mobile telephony. Yet humanity, thanks to what is probably Maxwell's most ingenious creation, has reached a level of interconnection unimaginable only a few decades ago. Human knowledge is now accessible at all times and from every location thanks to a simple portable device. Maxwell (but neither Hertz nor Marconi, who even transformed Maxwell's mathematics into working devices) could not manage to imagine that his mathematics would have such a great impact on human life. Maxwell's aim was simply to develop a theory which was logically perfect and his conviction was that, thanks to this theory, all known phenomena, but also a myriad of phenomena not yet known, would be understood and applied in technology. Thanks to Maxwell's electromagnetic wave theory and its developments, it is now possible to see inside a human body without dissecting it, send information anywhere at the speed of light, build systems to transform sunlight into electrical energy and much more. However for our mind, the importance of the physical phenomena related to electromagnetism is difficult to perceive. In fact, we do not have any direct sensitive experience of such phenomena.

Social phenomena have a clear impact on our life.

Instead, it is very immediate for each of us to understand the relevance, in our lives, of the mechanisms, which govern social life. Everyone understands that it is important to study the reasons why a skilful use of social networks can change the outcome of important elections and many people wonder how it is possible that a young opinion maker can earn millions of euros simply by broadcasting «live» his/her life, including some very intimate moments. No one is surprised at this difference: while rarely does a human being have direct experience of electromagnetism, all human beings must be reasonably familiar with social dynamics, if they do not want to endanger their own survival. Moreover, social intelligence is not a dowry developed and observed only in humans, but it is shared with apes, i.e. the primates genetically closest to us. This last consideration is confirmed by the results of the most recent studies of the primatologists. As reported for example in Russon et al. (1998) and in the essays of de Waal, our species, together with gorillas, chimpanzees and bonobos, has evolved, for at least ten million years, thanks to the development of a complex structure of social relations. Our social intelligence is the result of a long evolutionary process, which has left a deep trace on our genetic heritage. On the other hand, the appearance of human rational intelligence is certainly more recent. We will not deal here with the controversy that arises if we try to establish when, how and why this additional talent has been added to our genotype. Those interested can begin their personal investigations on this subject by reading Miller (2000). Here we will refrain completely from trying to discuss how much of our intelligence is shared with non-human primates (see again the monumental divulgation works by de Waal).

Dictators or Opinion Leaders?

There is one observation that is most frequently made by anyone who studies the process of forming a social choice, in any type of small or large group.

This observation reports (systematically!) the inevitable affirmation of the opinion of a particular individual, who plays the role of leader, guide, chief or, if you like, «dictator».

In Appendix 1, we present the history of the word «dictator» and the reasons why we prefer to use it instead of the other words. The reader is warned immediately that we use the word «dictator» with its specific original meaning in the Latin language. Let us anticipate it here: by dictator, we mean simply an individual to whom the group leaves the last word in case of disagreement on the social choices to be formulated.

The question that arises spontaneously is: why does every social group end up relying on a dictator?

Indeed, every group of individuals who associate in some form of social collaboration is organised following an internal hierarchy. In the hen houses, we observe the establishment of an order of pecking among the hens, an order which is respected at the most critical moment of their social life, that is, when it is necessary to regulate the access to food. Observations similar to those made in poultry houses are repeated in very many situations and for almost all species. The concept of pecking order has been extended to groups of chimpanzees (as well as to groups of all primates) and also to every form of human society. Think for instance to the enormous efforts spent in long periods of human history to establish nobility ranks. In Middle Age Europe, nearly every political dispute was ruled on the basis of this particular «pecking» order, which, like in bonobo societies, was rather stable, as it usually depended on the noble family's history and ancestry.

Many anarchist philosophers have wished for a humanity free from any form of dictatorship: how come their dreams have always been broken at the impact with reality?

Why a dictator must always exist?

Arrow's theorem, based on the conjecture of Condorcet, provides us with a precise and rational answer to these questions.

Is It Possible to Find a Democratic Method for the Determination of Social Choices?

We want here to start the discussion of

The problem of the determination of the social choice.

It can be formulated as follows:

GIVEN *a set of options presented to a certain social group and given the set of choices among these options made by each member of the group.*
TO FIND *a method to determine THE MOST DEMOCRATIC SOCIAL CHOICE, that is, the choice, among the possible choices, which must guide the actions of the group as a whole.*
We require that the social choice be "**democratic**"; i.e., it must be an «acceptable compromise» between all the individual choices made by the members of the group.

This problem is of great importance for humankind. For example, let us suppose that we have to establish how the waste collection cycle should be organised in an Italian region. Obviously, there are different technical alternatives, different interests, legal and illegal, different environmental policies, with different scales of values. Some would like to push for a differentiated collection of wastes, some would like to organise an industrial cycle, and some would like to use their own land as a simple dumping ground to earn immediately a lot of money. The problem of the social choice, which we are putting forward here, takes on a particular form: which are the decisions to be taken at regional level? How to get a reasonable compromise among the choices that each citizen would take if he were the only one who decides? To believe that this choice has to be made by a small circle of technicians is certainly naive. Among the conceivable solutions, the most effective ones require, in order to be feasible, a reasonable consensus in a large percentage of citizens. It is clear, in fact, that trying to implement the most efficient solution, that of differentiated collection, requires the almost unanimous consent of the citizens and presupposes that it is possible to keep under control those few criminals who have a personal interest in mixing the waste patiently differentiated by all citizens, in order to make a profit on the use of dumping ground for undifferentiated waste.

It is useful to understand when, why and by whom the problem of social choice has been formulated and recognised as important.

In the wake of the French Revolution, and shortly before being guillotined by the Jacobins who had just taken power, the Marquis of Condorcet studied the problem of identifying the electoral and political system that best reflected all the aspirations for equality *(égalité)* among human beings that the Revolution wanted to achieve. The historical moment was full of intellectual ferments. The Revolution, imbued with rationalist ideologies, wanted to endow itself with the best schools of mathematics in the world to deal more effectively with those technical and social problems that the old regime had not wanted or been able to solve.

In that historical moment, the attribute "mathematical" was not used to denigrate or ridicule, as if it were a synonym of "unnecessarily complex" or worse.

The Revolution favoured the rigorous and scientific analysis of social problems, and the mathematical thought was unanimously recognised as the most suitable tool to obtain their solution. The figure and *curriculum studiorum* of the engineer and the modern scientist were conceived at that time: the basic scientific culture was then recognised as fundamental for the training of technicians and professionals but also for educating those intellectuals destined to become politicians.

A Question of Rules or, Mathematically Speaking, a Question of Algorithms

Condorcet, immersed in his epoch cultural environment, decided to look for, by using mathematical tools, a democratic method for governing human societies. He understood that the most important feature of democracy is the establishment of rules, which are universally valid and certain.

Indeed, Condorcet knew (probably due to the works by Montesquieu, which we will discuss later) that there is a very important necessary condition for which a political system can be said to be democratic.

This condition requires that:

The way in which all social choices are formed and adopted, on the basis of the choices of the citizens, must be certain.

This means that a well-determined formal procedure must be set up which leads to the determination of the chosen social choice via a series of specific intermediate steps whose formal procedures are clear and certain. Some jurists, in this context, talk about the «computability» of magistrates decisions. It is only the «legal certainty», «separation of powers» and the «impossibility of changing or adapting laws in contingent situations», which can make human societies stable and fair.

Therefore, Condorcet made more precise the formulation of the problem of the determination of social choice, as follows:

How can we «calculate» the best social choice starting from the set of all individual choices? The crucial idea to be understood here is: what it means «calculating»?

In mathematics, since at least the famous Arabic language treatise textbook by al-Khwarizmi (about 825 CE), the concept of «calculation procedure» has been intensively studied and developed. The Latin translation of al-Khwarizmi treatise (whose Latin title was *Algorithmo de Numero Indorum*) introduced to the Western world both the decimal positional numbering system and the concept of «Algorithm». The Latin used word for the concept is, actually, a Latin deformation of the name «al-Khwarizmi»!

We can try to give here an informal definition for it:

An algorithm is «a finite sequence of well-determined operational prescriptions leading to a final output, starting from well-specified class of inputs».

With this definition in mind, let us re-examine a little bit closely the problem of Condorcet:

Let us assume to know:

(i) **the possible options** among which the choice has to be settled;
(ii) **the set of individual choices of all citizens**, i.e. all the orders of preference among the possible options operated by each citizen (each of these orders is an individual choice).

We want to determine the «social choice function».

The social choice function is the rule (i.e. the mathematical algorithm), which has to be used to establish the order among the options (i.e. social choice) to be adopted as a whole by the group formed by all the citizens.

Democratic social choice functions.

Obviously, the algorithm that, given all the citizens' orders of preference, produces the order of preference to be adopted by the entire group is more or less democratic, depending on how much and in which manner it takes into account the opinions (order of preference) of a more or less large number of citizens.

In this chapter, we will try to convey the basic ideas of the mathematical study of the Condorcet problem in an intuitive and informal way. To the reader interested to understand, in a more precise and rigorous way, the simple mathematical ideas which allow for its solution we will suggest further suitable readings in the annotated bibliography.

After many attempts, Condorcet realised that he was not able to find an algorithm to determine the social choice in a truly equitable way, starting from the choices of individual citizens. In fact, with simple considerations, he was able to demonstrate that all the methods used up to that moment (and all those he could devise himself) led either to social choices that did not faithfully and fairly reflect the will of each voter or to resolutions so ambiguous that they could not be considered effective choices.

The intellectual efforts spend by Condorcet are not without any practical use.

On the contrary, in certain situations, the solution of Condorcet problem (finally obtainable only in a weaker form, in which the demand for total equivalence of the opinions of citizens at all stages of the deliberative process is weakened) is crucial for the survival of a given social group.

Considerations on the Practical Use of Possible Solutions of Condorcet Problem in Politics

Here, we want to avoid discussing about the political situation in which Italy has been permanently since the decline of the political structure of the Roman Empire. Indeed, the reader can find a tasty and detailed description of the vicissitudes that Italy has had to go through in the last two millennia in "The History of Italy", a monumental work by Indro Montanelli.

We prefer, instead, to shortly discuss the situation in which the French State had put itself during its Fourth Republic.

The Constitution of the French Fourth Republic.

In said constitution, an enormous attention was paid to the balance of powers, the search for the sharing of choices, the representativeness of all opinions and all trends present in the society at all levels of the legislative, decision-making and governance process.

Such a constitution risked to be fatal for France: during the first colonial crisis that set Vietnam on fire, the French government lost the confidence of the Parliament .during the last desperate negotiations with the head of the insurgents, who found himself without interlocutors for months, months which he spent in the Côte d'Azur as a tourist; during the even worse crisis in Algeria, France risked, indeed, the final collapse simply for the lack of an effective leadership.

General De Gaulle.

De Gaulle saved his country by managing to get a series of constitutional changes approved, changes which led, eventually, to the Fifth Republic. Obviously, many French political parties thought that De Gaulle had seriously attacked democracy, reducing, in his country, the possibility of debating among different opinions. In fact, De Gaulle introduced into his country's legal system the figure of a President with quasi-monarchic powers: however, this President is elected by universal suffrage by the citizens, and Parliament can greatly limit his powers. So if the French citizens elect a National Assembly in which the President is unable to form a parliamentary majority, he is forced into a «cohabitation», which considerably limits his decision-making capacity. However, once a President of the Fifth Republic has been elected, and for the limited period of his or her term of office, he or she often has the last word on the most important among the choices that the State must make.

The problem, which Condorcet had raised a few centuries earlier, was of great relevance in his own country during the Algerian crisis!

A First Informal Formulation of Arrow's Theorem

A few years after the Gaullist reform of the French constitution, the Anglo-Saxon mathematician Kenneth Joseph Arrow would have shown that Condorcet and De Gaulle were both right: it is not logically possible to base a democratic system on a social choice algorithm which is both perfectly representative and capable of effectively managing political contingencies.

Therefore, Condorcet's analysis (as made rigorous by Arrow) shows that it is not in a single optimal social choice function that we must seek the best realisation of a truly democratic system.

A consequence of Arrow's theorem is that we are faced with a logical impossibility:

If a social group must take a precise decision out of a certain fixed list of options, it may not be possible to take the opinions of all citizens equally into account.

Therefore, if a political system has to be able to take decisions in every situation, also when the choice of the members of the group is very divergent, a decision-making procedure must be invented in which some citizens (or group of citizens) have a greater influence on the final choice. In short, the pecking order is a logical necessity: it is naturally established in every social group because of natural selection, and therefore, it is wise that its establishment be forecast in any constitutional and legal system.

Arrow received the Nobel Prize for Economics in 1972 because of his

Theorem on the possible functions of social choice.

It proves mathematically that the search for a «simple and perfect» democratic method is doomed to failure. Trying to be more precise, this theorem states that:

There is no algorithm which produces a social choice using as input the choices of the members constituting a group and which is truly democratic.

Arrow's result, both for the novelty of the methods of investigation it introduced and for the practical importance of the problems it considers, had a great impact in the social sciences and modern mathematics: it can be considered, indeed, the starting point of a new, promising and very rich field of research in applied mathematics. It should be noted that Arrow's original contribution to modern science does not consist in having invented, for example, the Theory of Sets (created by Cantor and developed among others by Frege, Russell and Zermelo) nor in having developed a new method of analysis of mathematical problems, as instead had done Tarski, Arrow's master and perhaps one of the most inventive and original logician of the twentieth century. Tarski, among other novel concepts and methods, discovered a very useful characterisation of the concept of infinity.

Arrow simply found a magnificent application of mathematical theories developed by others: by solving the problem of Condorcet, he made clear the importance of the intellectual achievements of twentieth-century mathematics, and in particular, he showed how powerful, in social science applications, is Set Theory. Notwithstanding its apparent abstraction!

Remarkably, the Nobel Prize was awarded to him and not, for example, to Tarski. Indeed:

Abstract theories are considered important only when they are proven to have a strong impact on the world of applications.

Arrow's theorem has encouraged many scholars to apply the rigorous methods of mathematics (and in particular of Set Theory) to the study of social sciences and, more generally, has attracted the attention of a wide public of non-specialists.

This circumstance is also related to the fact that, in a very suggestive way (and perhaps indulging in some sensationalistic temptation), Arrow's theorem has often been renamed by its popularisers as the Dictator's Theorem.

The Mathematical Model for a Choice: Total Order Relations

The key concept of Arrow's Theorem is that of «total order relation».

A total order relation is the particular type of relation, which is the mathematical model used, in particular, to describe a choice among several options, as made by an individual or by a social group as a whole.

More generally, given a set, every time when one «determines an order» among the elements of the set he can say that an order relation is established in the set. This mental operation is common in many activities of human intelligence, and it seems that also in the mental activity of apes such an ordering capacity can be observed (see again the works by de Waal).

More formally, we say that **a total order relation** is established in a set when the following two conditions are verified:

(i) given each pair of elements in the set, it is possible to determine which element comes first and which element comes next or whether the two elements are equivalent;

(ii) if the x element comes before (or is equivalent to) y, and the y element comes before (or is equivalent to) z, then x comes before (or is equivalent to) z.

The reader, after having pondered the previous definition, will agree that the mathematical concept of total order is the precise formulation of the intuitive concept of choice.

Indeed, given a set of admissible options, **a choice** consists of one of the possible ways in which these options can be totally ordered. Totally here is a needed adverb! It means that, in a choice for every pair of options, it is always possible to say if one option is preferred to the other or if the two options are equivalent.

Arrow's Rational Choice.

Arrow considers a particular class of choices in the specified set of options. He considers, in his hypotheses, only what he calls rational choices. There is an important nominalistic issue, which needs to be raised now. Words in natural language (as English language which are using here) have a «cloud» of meanings, all close one to the others, but still different. Instead mathematics needs very sharp meanings, which need to be made precise by suitable definitions. In facts, the adjective **rational,** as we use it when followed by the noun **choice**, has a specific meaning in the context of Arrow's theorem. Roughly speaking, it means that:

In the considered total order relation, every subset of options includes a most preferred one.

For making our discussion simpler, now on, we will often skip the adjective **rational**, when talking about choices, in the sense of Arrow. The reader must be aware of the fact that other theories may be formulated in which the concept of «choice» may be more general, and in these contexts, the use of adjectives like «rational» may be important.

Very often, in popularising essays, a rhetoric figure is used: as every choice is rational (can we imagine an «irrational» choice?), then the adjective and the corresponding hypothesis is not explicitly recalled. Some authors blame Arrow because he did use a «misleading» adjective. Following the Epicurean nominalistic attitude (so wonderfully also used by Archimedes), instead, we believe that one has to talk about «Arrow's rational choice» or «rational choice following Arrow», of course adding the precise meaning of the expression as defined by Arrow. Unfortunately, one has only a limited number of words in natural language and only seldom the invention of new words is justified.

A Second, More Precise, Formulation of Arrow's Theorem

We call **function of social choice**, the algorithm which produces, as an output, the social choice (i.e. the chosen rational total order among the possible options for a certain social group) using as inputs the set constituted by all individual rational choices (each individual choice being the total order among the possible options chosen by each elector).

Arrow's theorem on the Possible Functions of Social Choice may be now stated more precisely as follows

There is no possible function of social choice, which depends faithfully on the choice of all individuals belonging to the electoral body and is «democratic».

Arrow's theorem also specifies which is the actual structure of faithful functions of social choice.

In facts, the functions of social choice, which «depends faithfully» on the choice of all individuals, are either partially predetermined by "a priori" choices or are intrinsically dominated by a single individual (who is called «the dictator»). In other words, they are not democratic.

Of course, the previous statement must be completed by making explicit the meaning of the expression: «to depend faithfully on each individual choice». Following Arrow:

A function of social choice is said to «depend faithfully» on each individual choice if it is «monotonous and independent of irrelevant choices».

Now, the reader could be very upset. He could say that we are playing with words. We have indeed replaced the expression «to depend faithfully on each individual choice» with the expression «monotonous and independent of irrelevant choices». Both of them were not defined, and therefore, the reader is left only with the meaning of used words in English Language. Of course, these meanings are helpful, but many misunderstandings are still possible. Therefore, we need to give some further explanations here.

Arrow used the expression «to depend faithfully on each individual choice» to try to be understood by the experts of political and juridical sciences. Moreover, the term «faithful» conveys very democratic feelings.

While the word *democratic* refers to the fact that every elector is treated equally, the word *faithful* refers to the fact that the will of every elector is reflected in the social choice.

Arrow aimed to mathematical precision

So he based his study on the Theory of Sets and, in particular, on the definition of total order. Then, he precisely defined the concepts of monotonicity and independence of irrelevant choices. As declared few lines before, faithfulness is, in the sense given to these words by Arrow, equivalent to monotonicity and independence of irrelevant order.

Monotonicity and Independence of Irrelevant Choices

Here, we give some informal definitions. A more precise version of them will be given in the second appendix below.

What does it mean that a function of choice is monotonous?

If only a single elector changes his choice by increasing his preference for the option s while all the other voters maintain unaltered their scale of preferences, then the function of social choice can only increase the preference, in the social choice, for the option s, or at most must leave it unchanged.

We clearly expect that the previous property must be, indeed, verified by a faithful function of choice!

What does it mean that a function of choice is independent of irrelevant choices?

It is more difficult to paraphrase, without using mathematical formalism, this second definition. But: let us try!

Let us consider a subset T in the set of all options. Let us assume that voters change their choices about the other options while keeping their choices about the options in T unchanged. The function of choice is independent of irrelevant choices if the relative preferences, given by the social choice, for the options in T do not consequently change.

The most careful reader may still not be completely satisfied by the precision of the previously presented definitions. We are aware of this possibility. However, we are also aware that it is not possible to say what exactly the expressions "precise ordering", "depend faithfully", "democratic", "partially predetermined" and "intrinsically dominated" mean without making use explicitly of the formalism of the set theory. The interested reader will find the most rigorous formulation of the Dictator's Theorem in [Arrow] and in [Dardanoni]. In our opinion, these beautiful works are worth to be perused and understood. Finally, we tried to give more details in Appendix 2 of this book.

In particular, we believe that it is crucial to underline here the importance the concept of order relation. In fact, the human brain works very often by establishing relationships between objects, events or individuals. Children learn very soon to distinguish between some events, which happened before and other events, which happened later. This distinction is fundamental especially when one wants to establish a cause–effect relationship: clearly only what happened before can be the cause of a subsequent event! Children learn also very soon to say that something is bigger than something else or longer or heavier or stronger. The capacity of counting using natural numbers (one, two, three, four and so on) is strictly related, and probably co-evolved, with the capacity of ordering objects and is mastered by our children when they are about five years old.

In order to communicate or to solve simple problems of everyday life, we all use, sometimes consciously but more often unconsciously, the concept of relation of order: therefore, a full understanding of the related mental mechanisms is very useful.

Once more, it is possible to remark that our species is not the only one able to understand intuitively the abstract concepts we are talking about (see again de Waal). Indeed, many tests performed by different research groups show that chimpanzees can order in a sequence the first nine natural numbers. Moreover, apes can order objects based on their dimension and show clear preferences for objects, food, other apes and human beings.

However, it is reasonably certain that humans are the only living beings on earth, who can, consciously, formalise the concept of relation of order by using mathematical concepts.

Human beings should be proud of this unique capacity and cultivate it.

Arrow's Theorem Is a Powerful Tool for Understanding Many Social Phenomena

Anyone who has participated in the decision-making process in an academic or in a political body has "practically" understood the meaning of Arrow's theorem. Indeed millennial experience shows that it is not possible to find a set of rules ensuring that such a body could always manage to deliberate the needed decisions and, at the same time, that such decisions, always fairly and faithfully, reflect the views of all its members. Actually, in those situations in which each member of the body is able to put forward his or her own point of view equally, then all too often the collegial decision body fails to formulate a rational choice (as defined by Arrow).

Leaders of the Herds and Presidents

In order for a collegial body to function, it had to be accepted that its functioning be governed by a President.

Since the first modern Constitution, which was written by Pasquale Paoli for the Corsican Republic, the rules of operation of each collegial body provide in great detail the manner of election of its President. Moreover, the office of President, once he has been elected, always has a precise duration and some prerogatives, which cannot be anymore modified by the electing collegial body. In short, the President of the Senate or the Chamber of Deputies, for example, at the same time has the last word on the agenda of the Chamber that she/he presides over and cannot be dismissed, once she/he has been elected, by anyone, even by a vote of the same collegial body, which elected him/her.

Another absolutely uncontroversial experimental observation concerns the constant existence of a leader of the herd.

Given a herd of individuals of any species of animal (and also a human collegial body belongs to this general class of social groups), it has been always observed the existence of an individual, selected by all herds in very similar ways, who has the duty of making the decisions necessary for the survival of the herd when these decisions cannot be taken by taking faithfully into account the wishes of all its members.

© Springer Nature Singapore Pte Ltd. 2019
F. dell'Isola, *Big-(Wo)men, Tyrants, Chiefs, Dictators, Emperors and Presidents*,
https://doi.org/10.1007/978-981-13-9479-9_2

It has been noted that the leader of the herd has a relevant role in, at least, two types of different situations:

(i)   when the decision must be taken in a short time and it is not possible for some reason to consult all the individuals of the herd;
(ii)   when a decision is really necessary, although opinions in the group are so discordant that no sufficiently shared decision is possible (Italy, in its history, has given endless examples of such deadlocks!)

It is easy to give examples of situations described in the previous point (i): think of a leader of a group of chimpanzees who must decide whether to face the fight or whether to flee in the presence of aggression by an enemy troop, or think of President Kennedy during the Cuban crisis. Both the super-alpha males considered here had to make decisions in a short period of time and could not democratically consult the group they were leading.

The office of dictator was conceived by the first Roman constitutionalists in order to give the Roman Republic the capacity to react effectively to those unforeseen contingencies that require immediate reactions: but, if initially, the position of dictator was limited in time, Caesar managed to have it assigned permanently, thus distorting the mechanism of functioning of the Roman republican institutions.

Arrow's theorem can be applied successfully to describe situations that fall into the second class considered in point (ii): those in which a group does not manage to make decisions because of the impossibility of reaching an agreement among its members.

Arrow's Theorem and the Selection of Super-Alpha Individuals

Arrow's theorem provides also a good conceptual basis for the rational explanation of the existence, in every social group, of a super-alpha individual.

A super-alpha individual can be defined as an individual who has the last word in all situations of stalemate that occur in the internal decision-making process of the group which he is dominating.

The analysis of the phenomenon "existence of a super-alpha individual" can be sketched with the following flow of statements:

(1)   Experimental evidence points out that there is a leader in every group of individuals of every social species, that is, in every species in which the survival of an isolated individual is impossible.
(2)   A system of rules realising the «perfect» democracy has not been found up to now. The utopia of a system of egalitarian rules leading to the formation of the choice of a group as a whole, starting from those of its individuals, all treated as «peers», has been realised neither by philosophers and mathematicians nor

by the process of natural selection. Instead, every political system and natural selection observed up to now have invariably produced a pecking order in groups dominated by a super-alpha individual.

(3) Clearly, in the natural competition between species, the advantage of living in groups is so great that many species have adapted to have a social life.

(4) Those individuals who belonged to groups led by a super-alpha individual (rather than left in a situation of anarchy) had a better chance of surviving.

(5) Very soon all social species were formed by individuals genetically adapted to accept that group life be regulated by a dominant individual.

Arrow's theorem assures that one should not be surprised if the search described in point (2) did not produce any result.

Indeed the arguments developed by Arrow show that there is a logical impossibility, which makes difficult the formulation of democratic rules for governing decision-making bodies.

Arrow's Theorem as a Tool for Understanding History

Many Jacobin revolutionaries have wondered how it has been possible that throughout human history, all societies have accepted the existence of dominant individuals and why all societies have developed a highly hierarchical structure divided into classes.

The Jacobin analysis of history is extremely simplistic. Indeed they believed that human nature is schizophrenic. For this reason, they concluded that, in human history, a majority of individuals, psychologically weak, have always been exploited by a minority of violent and arrogant individuals.

Perhaps the most extremist Jacobins lacked the culture needed to appreciate the analysis attempted by Condorcet and, perhaps also for this reason they killed him, after a summary trial. Their analysis of the reasons for human behaviour was too rudimental and could not be compared, in depth and quality, with the analysis developed by Condorcet. As too often happens, in these situations, Jacobins, to avoid to consider deep, very advanced and sophisticated concepts, and believing that their practical approach was more efficient, decided to use a «Stalinist» solution to solve the political controversies started by Condorcet and preferred to get rid of him physically.

Jacobins (and later Stalinists), after having blamed the citizens who accept the dictatorial decisions of a minority of arrogant individuals, decided that their minority group was right and imposed their dictatorial decisions to everybody else!

Human nature is not schizophrenic at all.

Condorcet-Arrow analysis has indeed proved that the presence, in a group, of dominant individuals is a logical necessity!

No group can survive without an individual, i.e. a dictator, a President, a king, etc. (or if the group is large some individuals, i.e. a class of nobles in some societies, a group of officers in the army, etc.) which are selected among its members and whose task is to resolve disputes and to lead the group to make decisions in all circumstances, even in absence of opinions accepted by a sufficient majority of group members.

The pecking order, that is the hierarchical order established in a given group by means of ritualised physical clashes (clashes which in some groups of intellectuals, are called exams or competitions for getting a university chair), allows the group to establish a priori who should have the last word in situations where there is not a sufficiently shared agreement in the group.

The advantage for a social species in establishing a ritual to form an internal hierarchy, and in regulating its own life by entrusting the solution of controversial situations to a dictator, is really very relevant. Consequently, many of the aspects of the behaviour of individuals in social species that concern the formation and maintenance of an internal hierarchy within their groups have been, most likely, fixed in parts of their genetic code.

Probably for this reason, the Jacobins did not succeed in eradicating from the minds of their fellow citizens either the religious sense, or the respect for the hierarchies. They did not even succeed in changing the calendar, dividing the months into decades instead of weeks. The failure of the Jacobin political programme was inevitable as their reforms, too often, went against human nature.

All social species have a genetic background that leads each individual to respect the hierarchy and, at the same time, to participate in the competition to establish the order of pecking following a predetermined rite.

There is no schizophrenia in human nature: all individuals are adapted to optimise their survival possibilities. If the selected leaders are in power, they respect the pecking order, and if there is a competition, then they are ready to fight to get some power.

With regard to the reform of the calendar, Jacobins' failure was even more easily predictable, because the theoretical knowledge necessary to predict failure was completely available to revolutionaries. Indeed the needed theory is not so complex to require difficult reasonings as those involved in the proof of Arrow's theorem: the menstrual cycle of women has a period that is a multiple of seven days, and the traditional calendar, which Jacobins tried to reform, is organised on a weekly partition. While it is very rational (and natural) to divide the measures of every quantity into submultiples using the basis ten (humans have ten fingers!), it was impossible to manage to impose a calendar in which months were subdivided into decades. A feminist analysis of the latter issue could lead to the conclusion that many of the Jacobin failures were caused by the fact that they never wanted to consider women as full citizens.

Arrow's Theorem as the First Mathematical Result in Psychohistory

Actually, we believe that a rational scientific analysis will shed new light on many of the historical events that have been handed down to us. This analysis will be possible using, for instance, evolutionary ethology (see [Wade]) and the many other theories developed, recently, by applying the scientific method to the description of social phenomena.

Many will consider the previous statement at least as reckless, others will reject it sharply, and some others may believe that the scientific theories now available are, still, very far from being able to give accurate predictions and descriptions of social phenomena.

This is not the opinion championed by Isaac Asimov. He is, in our opinion, one of the greatest thinkers of the twentieth century. He, with his work of popular scientific fiction, gave a major contribution to raise the knowledge of large strata of our society: Asimov was sure that the scientific analysis of the huge mass of social phenomena that we call "history" will be possible in a not too distant future.

Everyone should have the wonderful experience of immersing himself in the imaginary world of Asimov.

Many decades before these scientific achievements he was forecasting became actually a reality, he considered as feasible: electronic computers, Internet, computer controlled image projector, space navigation and a myriad of other technological innovations. Indeed he imagined them with great precision: his description of these devices is so realistic that many readers, in 2018, do not realise that his books were written in the fifties (of twentieth century!).

Asimov's Psychohistory.

Well, Asimov's most fascinating vision concerns the future formulation of a revolutionary Theory, which he calls *Psychohistory*. Psychohistory should allow us to predict the behaviour of large human social groups: it is applied by the protagonists of Asimov's stories both to explain past events and to predict future ones.

We believe that Arrow's theorem is a very first step towards the formulation of some form of psychohistory.

Some may think that Asimov is just a lucky science fiction writer who is granted suggestive «poetic licenses» but whose imagination is far from reality. In fact, Asimov is much more than a visionary intellectual: being an associate professor in biology at a prestigious American university he soon decided to devote himself to popular science literature. His unique gift of synthesis in writing led him to produce magnificent introductory texts to physics and chemistry and his activity as an author of science fiction novels can be considered the ideal continuation of his activity as a populariser and inventor of realistic future scenarios.

One should think, for example, of "Robotics". Today, it is the name of a serious and very practiced science: there are thousands of professors of robotics in universities around the world. The word robotics itself, together with a very precise definition of its field of application, were both conceived by the mind of Isaac Asimov as he wrote his books of «science fiction».

By the way, the word "Robot" seems to have been used the first time as a word for meaning a "working artificial automaton" in the play R.U.R. by the Czech writer Karel Čapek. Karel, however, attributes to his brother Josef the invention of the word robot, with the specified meaning. It has to be remarked that in Slavic languages «robota» means a «forced laborer», of course referring to human beings, however.

In any case, by imaging the development of future technology, Asimov managed to outline a precise and fruitful field of research.

Another impressive proof of the predictive abilities of Asimov's mind is found in one of his first novels describing the imaginary process of the birth of psychohistory. Hari Seldon, the imaginary mathematical inventor of psychohistory, arrives in the capital of the Galactic Empire and, not knowing the places, asks for help to arrive at his hotel. He is given a tool with a viewer that talks to him and shows a map that guides him to destination. More than 50 years before its actual invention Asimov describes, in great detail, a TomTom, i.e. a device produced in twenty first century.

"A Priori" Imposed Social Choices, Alternatives Which May Be Relevant, and Institutional Systems Incapable to Decide

Many further interesting considerations are made possible by the increased descriptive power gained, because of Arrow's theorem, by the rational analysis of historical events and social dynamics. We select here some among those which we considered more meaningful.

As repeating them in different forms is useful for improving understanding, let us once again reformulate the basic concepts in the enunciation of Arrow's theorem.

1. A function of social choice associates, algorithmically, to the rational choices of all involved individuals a precise rational choice, the one, which should be adopted by the whole group.
2. Arrow definition of «democratic» function of social choice requires that the function of choice must be independent of irrelevant alternatives and must be monotonous: seen the meaning of these assumptions, as discusser before, it seems very difficult to give them up!
3. A function of social choice that verifies the hypotheses of Arrow's theorem must necessarily be dictatorial (i.e. it cannot make all the voters weigh equally) or it must have choices imposed a priori.

Concerning point (1) it should be noted that a social choice function is necessary to produce a group decision as a whole. Recall that, by definition, the function of social choice is an algorithm, i.e. a rule or an automatic procedure, which, considering all the choices of the single individuals, mechanically produces the choice that the group must adopt. To assume that a function of social choice can be determined is equivalent to require that the group is always able to make decisions. To assume the existence of an algorithm which can be used without ambiguity is the essence of the affirmation of the «certitude of law», in the decision-making process.

Now it is evident that, in human advanced societies, such an algorithm has not always been established. Often, as in France during the Fourth Republic or in Italy, nearly always in its history since the fall of Roman Empire, it has been preferred not to make decisions rather than not to take into account the opinion of some, even relatively small, subsets of the social group.

However, the most astonishing case of «blocked system» (i.e. a political system incapable to make decisions) whose existence has been documented is the one, which was developed in Polish-Lithuanian Commonwealth. We are talking about the infamous «Polish Veto» also called «liberum veto» (a Latin expression meaning "a veto which can be freely expressed"). It was a unanimity voting rule which allowed to any single member of the Commonwealth legislature (in Polish «Sejm») to veto a single law, in the first period of the application of the rule, or to veto the whole deliberations of a legislative session, in a second period. The rule was based on the principle that all Polish noblemen were equal. While the «liberum veto» did have the effect of increasing the democratic nature of Polish institutions by balancing and checking the royal power, it eventually caused the dissolution of the Polish-Lithuanian Commonwealth, becoming the cause of the famous "old Polish anarchy". Finally, the expression «Polish parliament» in several languages became a synonym of «a political situation in total paralysis».

Arrow's theorem clearly indicates which is the first countermeasure to use to remedy the logical impossibility it demonstrates: if we want to maintain the demand for equality in the weight of the choices of every voter then we must give up part of the ability to make decisions. Of course, the Polish Parliament was a too extreme countermeasure, indeed.

On the other hand, Arrow's theorem does not allow for an analysis of the loss of efficiency of a social group, which is paralysed by the inability to take some among the decisions which are really necessary. We refer, as an example, to the decision of cutting public spending to reduce an enormous public debt accumulated by corrupted political administrators.

Indeed very often not making decisions is more dangerous than making a wrong decision and therefore a mechanism which blocks decisions in the absence of a strong agreement can be very detrimental for the society well-being.

The mathematical analysis of the effects of blockage on the development and survival of societies seems very important and we hope that some mathematical economist may be willing to dedicate his life to develop it.

With regard to point (2) it should be noted that the demand for independence of irrelevant alternatives is not always appropriate in the dynamics of social groups. A democratic system can decide to change its order of preference in a certain subset of options if it can induce favourable changes in the resulting social choice regarding other options. In short, if the vote to have a road built instead of a bridge will make preferable to finance a school instead of a hospital, then the road may become preferable to the bridge. This type of situations probably deserves further analysis.

With regard to point (3), it should be noted that there are countless examples, in history, of situations in which it is very difficult to establish if a social choice functions is «a priori imposed» or if it is «dictatorial». Moreover, there are many situations which one can recognise to be partially a priori imposed and partially dictatorial. Some taboos or religious prescriptions seem to have been developed simply to get out of the indeterminacy in the social choice, as described by Arrow's theorem. Indeed it is sometimes difficult to distinguish clearly, when examining the structure of taboos, if they are imposed or if they are dictatorial choices.

One can recall, for instance, that the resolution of hereditary disputes between German princes of the Holy Roman Empire was often referred to a Papal arbitration, as it was claimed that the Pope was pronouncing the verdict not as his dictatorial choice, but simply as a choice imposed by the higher divine will.

In fact, it is often very difficult, in practical social instances, to determine whether a certain function of choice is imposed or if it is dictatorial. Let us examine this point more closely.

A function of choice is said to be imposed, when it is based on a set of choices that all members of the group accept as indisputable.

For example, having to choose among the candidates for a certain elective office, it has been established that if some candidates have obtained the same number of votes, the candidate to be elected is the elder one. Here is a shared value (in human groups but also in bonobo groups) that solves some situations in which the mechanisms of social choice could be blocked. Similarly, the set of values imposed by a religion serves very often to reduce the situations of stalemate in the process of forming social choices.

Most likely, the most frequent criterion, for imposing choices in the functions of social choice, which has been observed is that based on age and we often hear of political systems based on gerontocracy, that is, the government of the elderly: well, this particular way of selecting the dictators necessary to make any political system work has very often been adopted in human history. For example, the Gerousia was the collegial body that governed political life in Sparta: to become its member one had to be sixty years old, an age that at the time was reached only by a few lucky ones.

Even in groups of primates, the highest seniority is considered in many cases as a criterion for selecting dictators and for establishing the hierarchy within the group. Think of the groups of bonobos: in them, social life is regulated by a matriarchal system and for this reason, violence is almost completely unknown. Each social conflict is solved, by bonobos, in a ritual way and each of the social rites of the bonobos involves, in a relevant way, sex. In bonobos, sex has been transformed into an instrument of cohesion of social groups. Actually, when in a group of bonobos, it is necessary to establish the order of pecking then the females place themselves in a hierarchy, which sees at the top the oldest ones, independently of their capacities. In the other hand, males are placed in a hierarchy, which reflects that of the respective mothers. It must be said that the ethologists who have studied the societies of bonobo have compared them to a paradise of peace.

The difference with the violent, competitive and sometimes ferocious societies of chimpanzees or gorillas is amazing, if one considers that these three species are genetically very close one to the other and also to humankind. However, it should also be noted that the bonobos show a very low ability to face the competition for survival compared to other groups of social animals: they seem to succumb too often, precisely because of their inability to violent behaviour. In their societies, competition between individuals is relatively very low, since the main criterion imposed to determine the weight of individuals in the decision-making process is that of seniority.

Evidently, the social mechanisms that have forged the society of the bonobos have also been triggered in the societies formed by Italian university professors.

In the latter societies, the most common criterion in establishing hierarchies is the seniority in the role (which can sometimes be different from age) or the power of the professors' (academic or genetic) family. The electoral system for the choice of selection committees in Italian universities provides that, in the event of a tie between the votes obtained by the candidates, the eldest-in-the-office-of-professor is elected, and in the event of equal seniority in the office (which is not too infrequent given that Italian professors have often been recruited simultaneously by national competitions), the-eldest-in-age is the elected one. Moreover, anyone who has even tried to start a university career in Italy has been explained by his academic mentor that, for the attribution of the positions of professor, the preferred criterion is seniority, i.e. the number of years of non-tenured employment or permanence in the professorial rank lower than that which one aspires to occupy. This criterion is often justified stating that it is often objectively very difficult to determine who really has the greatest merits.

Just like the bonobos' groups, the Italian universities have established that, if the function of social choice they have given themselves is not able to produce a rational choice starting from the rational choices of the professors in office, then the choice imposed to be used is that of favouring seniority.

On the other hand, a function of choice is said dictatorial if it appeals to the judgment of an individual to resolve disputes.

The reader will think about the history of Western countries and will consider how difficult, if not impossible, it is, in the case of societies in which the Catholic Church operates, to establish whether a social choice is imposed by God (i.e. by a function of choice imposed a priori by shared values) or if, instead, it is dominated by a dictator (the Bishop of a Diocese or the Pope). It should be also noted that the Protestant Churches ideologically founded their schism from the Church of Rome precisely on the distinction between social choice imposed by values (as established once forever by God) and that imposed by the «dictating» bishop. When Luther affirms that each Christian must make his decisions by consulting the Bible and putting his conscience in contact with God, he actually affirms that the set of values that guides the imposed social choices are already present in every man and that it is not necessary to recur to an authority to reaffirm the moral dictates. In Luther vision a dictator who imposes an authentic interpretation of the scriptures and directly guides the group towards a social choice is useless. On the other hand, it seems that the Church of Rome does not have the same confidence in the conscience of the individual: it believes that the guidance of the choices of individuals must be constantly monitored by the hierarchy, even through the periodic confession of every believer. The reader should form his own opinion about the relative efficiency of the different Western social groups according to the religious beliefs they have embraced: was it more convenient to adopt the anarchy of Protestant values or the uniformity of judgement of Catholic Church? Finally, it should be noted that more recently, Pius XII strengthened the dictatorial nature of the function of social choice in groups where the Catholic faith is more widespread. Actually, Pius XII affirmed, talking *ex cathedra*, the dogma stating that the Pope is infallible, when he talks *ex cathedra*.

We feel to give here some few further explanations, to the readers who are not expert in Catholic Dogmatics. Papal infallibility is a dogma of the Catholic Church. It states that the Pope is protected by Divine Providence from any possibility of error *«when, in the exercise of his office as shepherd and teacher of all Christians, in virtue of his supreme apostolic authority,* (i.e. when he speaks **ex cathedra**) *he defines a doctrine concerning faith or morals to be held by the whole Church»*. This doctrine was explicitly stated during the First Ecumenical Council of the Vatican (1869–1870) but had been considered implicit by all Catholic medieval theology and was shared by the majority opinion of Catholic hierarchy at the age of the Counter-Reformation.

In Catholic theology, papal infallibility (**i.e. a form of power of dictatorial choice**, following the definition by Arrow) is needed to solve the controversies among different theological interpretations of Sacred Scripture as acted by the Sacred Tradition and different Bishops. However, in the documents establishing the Dogma of Papal infallibility, it is explicitly stated that «the infallible teachings of the Pope must be based on, or at least not contradict, the Sacred Tradition and the Sacred Scripture». In other words, this Dogma states that the dictatorial power of the Pope must be based on **a priori imposed choices**: those deriving from the Sacred Scripture. As the extensor of the Dogma was aware of possible situations of «decision blockage» he added the expression «or at least not contradict». However, it is not stated, in the Dogma, how the implicitly needed check of non-contradiction between Papal teachings and the Sacred Scriptures is performed. Therefore, one can conclude that the Constitutional system of Roman Catholic Church has some flaws.

The reader having some knowledge of the basic ideas of mathematical logic will recognise that when a Pope, speaking **ex cathedra**, states that he is infallible when speaking **ex cathedra**, creates a kind of paradoxical situation recalling very much the one found in the antinomy of the liar.

King, Emperors, (Absolute) Monarchs and Despots

The social structures observed in herds, troops, flocks and primates societies have been determined by natural selection. Humans, instead, at least since the raise of one of the most ancient State in history (we think of Babylon), have tried to establish some rules for regulating the social life (Hammurabi code!) by the use of rationality and common sense wisdom. The reader, however, should not underestimate the results of the optimisation process, which has shaped primate's societies and human nature under the push of natural selection. The forms of primates (and in particular humans) social organisations shaped by the long-lasting Darwinian evolutionary process are, indeed, very advanced and may inspire the formulation of the rational theory of constitutions. In particular, we believe, the process leading to the selection of a dictator deserves more extensive ethological and theoretical studies: in this aspect the available literature seems rather poor.

Hereditariness of the dictator role.

In the human species, probably in order to reduce the struggles necessary for the selection of the dominant individual or dominant group of individuals, the principle of hereditary attribution of the role of the dictator has been affirmed.

In many social groups, some families have won the right to pass on the role of the dominant individual, from one to the next generation, to the first-born male, and, in case of lack of male heirs, also to the first-born females. Unfortunately, although it is clear that children inherit many of the physical and psychological characteristics of their parents, the talents of intelligence, balance, cynicism and ferocity necessary to exercise the role of the dictator are not inherited with exactly those laws of heredity, which were discovered by Mendel for some phenotypes of cultivated peas. Unfortunately, the ability to be a good dictator is not like the colour of the eyes!

In primates, the ability of being a good dictator is such a complex feature that it has been observed rather seldom, in human history, that the son of a skilful King has been indeed as skilful as his father.

We, therefore, conjecture that the rule of hereditary transmission of dictatorship role is rather to be related to a kind of «innate tradition», as the one observed in bonobo troops. Bonobo and human societies seem to have proven that a rule whatsoever for transmitting power without violence has a positive impact on the survival of a society. Bonobo societies have established a pecking order based on age of the individuals, humans on the transmission of nobility titles from father to son.

An example of non-hereditariness of dictatorial skills: Frederick II Hohen-staufen and his successors.

An example of exceptionally gifted personality is given by Frederick II Hohenstaufen (26 December 1194–13 December 1250). He was King of Sicily from 1198, King of Germany from 1212, King of Italy and Holy Roman Emperor from 1220 and King of Jerusalem from 1225. Under his reign, the Holy Roman Empire attained its greatest territorial extent ever. The great majority of historians have described him as an exceptionally clever, educated, wise and capable guidance of his people. For instance, Donald Detwiler wrote about him:

> A man of extraordinary culture, energy, and ability –called by a contemporary chronicler *stupor mundi* (the wonder of the world), by Nietzsche the first European, and by many historians the first modern ruler– Frederick established in Sicily and southern Italy something very much like a modern, centrally governed kingdom with an efficient bureaucracy.

However, his mother Constance (Queen of Sicily) did not show any particular inclination towards higher education or visionary politics, although she understood how to assure to her son the protection of the Pope, and she was clever enough to deliver him in public, after having heard gossips insinuating that she was too old to be really capable to be pregnant. His father Henry VI of the Hohenstaufen dynasty did not show any capacities as an Enlightened Monarch and behaved in such a violent way against his people that, most likely, in order to stop his bloody politics, his wife organised his murder.

On the contrary, Frederick political and cultural capacities were enormous: he transformed Palermo into a major European Capital, and from there, he managed to rule Sicily and the greatest part of Italy, up to all the way north in Germany. With a bloodless Crusade, he managed to ransom Jerusalem, becoming its king. His, undoubtedly great, genetic talents were cultivated by four preceptors: one of Latin culture, one Greek, one German and one Arab. As a result, he could fluently speak six languages (Latin, Sicilian, Middle High German, Langue d'oïl, Greek and Arabic), and was a munificent patron of the sciences and the arts. He also decided that in his kingdom there was the need to establish the certitude of law. In his enlightened codes, among many innovations, he declared the trial by ordeal as superstitious and cancelled it as a tool for dispensing justice.

However, Frederick never conceived the possibility to share his power. His ideology was simple and crystal clear: he was Emperor in the name of God. He claimed to join in his hands the temporal power of Caesar Augustus and the highest religious dignity. Useless to say, because of his pretensions, the Papacy started a violent fight against him: Frederick was excommunicated four times and systematically calumniated by pro-papal chronicles: Pope Gregory IX officially declared that he was the Antichrist.

Frederick's heirs, because of his great consideration for higher education, had the best possible preceptors. However, they were not as capable as he was, and his family shortly after his death lost completely all its dominions. The Papacy, being an institution based on rules conceived for lasting thought millennia eventually prevailed.

The life of Frederick is, alone, enough to establish how ineffective is Absolute Monarchy as a system for choosing the dictator of a society.

Even an outstanding personality, as he was, did not manage to stabilise in time the very efficient system of administration, which still he had been able to establish. On the contrary Papacy, in direct competition with the Sacred Roman Empire, showed that its constitution was more effective, and indeed it managed to rule Italy and a great part of Europe for a long while, after the collapse of the family Hohenstaufen.

Kings, Emperors and Monarchs.

Various nuances and great differences can be found in the role played by Kings, Emperors and Monarchs who acted as «dictators» in different societies and times. For instance, in [Viola] the pre-revolution French constitutional system is analysed:

the discoveries that this reading allows for are remarkable. The Absolute Monarch was not a Despot: his power was limited by customs, formal rules and even by the laws he himself had previously promulgated.

Do not think that a French (Absolute!) Monarch could easily promulgate laws: if the Paris Parliament did not want to accept a law, it could refuse to gather in its *quorum,* so refusing to hear the law pronounced by the King laying on his sacred bed of justice. In absence of the proper legal ceremonial, the law could not become effective: in other word the Parliament could veto the King decisions. Actually, this is the reason for which the last Kings of France could not solve the social tensions between Nobility, Clergy and Bourgeoisie. The nobility did not allow the Kings to do so! This situation had also an unexpected consequence: very often it happened that the King was forced to respect a law that he himself had promulgated as he was not able to modify it. Commenting on this situation, the members of the French Parliaments very often responded to the complaints of the King by reminding him that «Your Majesty, You are not a despot!».

Even in a group of chimpanzees (see [de Waal] and [Russon]), the super-alpha male must respect some rules and undergo some consensus procedures if he wants that his decisions be accepted by the group. Moreover, if one wants that human societies attain a state of peaceful prosperity, he has to forecast, as done by the Roman Republic, a constitutional frame in which the dictators are chosen and can act in the interest of their people. A constitutional frame must be conceived to avoid that the magnificent achievements obtained by the most gifted dictators may be lost, as happened to Frederick II Hohenstaufen.

In short, while a dictator for sure contributes to the forming of social choice in a more relevant way than other individuals, in general, he never manages to impose his will in an uncontrolled and indiscriminate way, and his actions, without a legal system preparing his succession, are bound to be vain.

In fact, to be an effective dictator is the result of a great quantity of hereditary characters, characters that are mixed up in a significant way in the passage from one generation to another. Therefore, the fact that a father was a good dictator increases in a negligible way the likelihood that his son will be a good dictator. On the contrary, the probability that the son of a good dictator is a good dictator himself is essentially equal to the probability that any other individual is a good dictator. This is the reason for infinite instability in the societies, which have delegated, in their evolution, the role of a dictator to the descendants of a man who had the necessary qualities for this role. Everybody who tried to understand the history of Europe, where the ideology of hereditary Monarchy has been more deeply rooted, has observed the endless sequence of changes of dominating families, of violent civil wars and of continuous murders and abuse of power which characterised its societies.

The list of the names of sons who were not capable to continue the political action of their father could fill many volumes.

We want simply recall here that the lack of a capable successor of Lorenzo il Magnifico as (*de facto*) Lord of Florence probably changed the destiny of Italian peninsula, abandoning it in the hands of foreigner European powers. The life of Lorenzo needs to be studied carefully. Exactly as Caesar, he never formally changed the republican nature of the State which he eventually managed to rule: instead, he could systematically impose his dictatorial point of view to the Republican collegial bodies charged of the most important decisions. Of course, the fact that he controlled one of the most powerful banks of his epoch did help a lot.

Democracy as a Possible Alternative

The Hereditary Monarchies has been considered, during several centuries, as an effective alternative to the system developed in ancient Greece: Democracy.

The role of dictator, in Democracies, is entrusted to individuals who undergo a selection process, the election campaign, which culminates in a formal election by all citizens. Each elective office has a limited duration and rigidly fixed prerogatives.

The opposition between these two political systems has lasted for millennia and evidently neither of them has such advantages as to have supplanted the other.

Actually, for a long period in Western Europe, the Greek Democracy has been spoken of as a failed attempt to organise society! Indeed Greek Democracy did not know the concept of separation of powers and the Popular Assembly, by advocating the highest sovereignty, became easily manipulated by populist manoeuvres.

One must, however recall that there are, as for the Monarchies, different forms of Democracy.

Since the function of perfectly democratic social choice is not logically possible, a democracy, in the meaning of the concept which has been developed after the American and French Revolution, must, as indicated by Arrow's theorem: (i) appeal to a set of values in order to obtain imposed choices, (ii) choose, in precisely defined cases where a decision is recognised as essential, one or more dictators by a procedure as shared as possible, and limit the dictator's powers by means of a system of balances and checks, (iii) accept to use a function of choice that it is not always capable to choose, so accepting, sometimes, to get stuck in the decision-making process.

Clearly, the constitutionalists or the founding fathers of one of our political systems chose different balances between these three different treatments to the logical impossibility shown by Condorcet and Arrow. In this way, many different possible democracies have been conceived.

The characteristics of societies resulting from different political systems can be very different. Reference will be made here only, and by way of example, to the enormous differences manifested by two political systems, which, in the broad sense that we have specified here, can be called democracies. We are talking about the Athenian democracy and the Spartan oligarchic system. As the first shows respect for the opinions of individual citizens, the second calls for a set of imposed values and its functioning is based on the diarchy of two kings. If Athens has produced a refined culture, fruitful and profound thinkers, some of the most important texts of Western philosophy and law, Sparta has managed to prevail in the Peloponnese war, war wanted by Athens and which marked Athens' political and economic end. The political system of Athens was too fragmented to withstand the confrontation in a direct clash with the rigid, but more solid, system that governed the Spartan institutions. The democratic Athens succumbs in the confrontation with the oligarchic Sparta. Remark that, instead, the democratic system established in USA did manage to prevail on the apparently more solid system established in the Soviet Union. The parallelism USA/Athens and URSS/Sparta revealed to be completely wrong: actually, the balance of powers intrinsic in the USA constitution made USA the most solid political system in our époque.

Indeed, the «formal» Roman democracy represents a remarkable improvement of the populist Athenian democracy: in Rome, the management of the crises is faced by means of the election of a dictator and the system of control of the elective magistracies is very effective. The duration in office of a dictator, until the imperial period, is strictly limited to periods of crisis, and the consuls, the magistrates who hold most of the executive power, last in office only one year, form a dual magistracy and alternate, exchanging roles every day at the command of the army. The outgoing consuls preside over the electoral meeting that elect the new consuls: the respect of the formal procedures is so careful that immediately after the terrible defeat of Canne, due to Hannibal, the electoral meetings were summoned and the elections were celebrated as if nothing had happened.

The great efficiency of the Roman political system was the first cause of the dominion that Rome exercised over the whole Western world, a political dominion that succeeded in subduing Hellenistic States which, on the other hand, had managed to reach peaks of scientific and technological development unequalled until the end of the eighteenth century by European States. In the confrontation between highly technologically and scientifically advanced Hellenistic societies, however, weakened by their inefficient political organisation, and the Roman relatively underdeveloped society, which, however, had managed to organise its institutions in a very effective way, it was Rome to prevail.

The Roman political system was a masterpiece in balancing between the capacity to make decisions, which was assured by the presence in that system of monocratic magistracies, and respect for the will of the electorate.

The ability to act in situations of crisis or in the presence of widespread disagreement in the electorate was ensured by the efficiency of the electoral system, which for a long period managed to select highly qualified magistrates, to whom were granted the prerogative to act for the best. On the other hand, the respect for the will of the electoral body was ensured by an apparatus of formal rules, separation of powers, constitutional guarantees and deep-rooted democratic traditions: Roman citizens always felt protected by their republican institutions.

In more recent times, we must remember the tragic experience represented by the Republic of Weimar. In Germany, between the two great wars, a democratic system, by hyper-guaranteeing the will of all minorities, paralysed the decision-making process and led to the democratic election to the state highest executive office of the tyrant *par excellence* of the twentieth century: Adolf Hitler. A social group, tired of the inability to make decisions of a political system that cared too much about the balance of powers and respect for the choices of individuals, eventually handed itself over to a mad despot.

More effective solutions to the problem of the balance between decision-making capacity and the representativeness of society's opinions are given by two of the most efficient Western democracies: the US Presidential Republic and the Fifth French Republic.

The American case is an example of the realisation of the equation "president equals dictator for a specific time period" while the French case is considered an example of the equation "president equals constitutional monarch for a specific time period". The temporary-dictator-USA-President forms a government that is not subject to a vote of confidence by Parliament, cannot hold office for more than eight years, must be re-elected after the first four years term, can spend money and levy taxes only if Congress allows it, can veto a law passed by Congress and is obliged to promulgate it if the Congress votes it again with a qualified majority. The American system, conceived at a time when a non-monarchical political system was considered impossible, has shown that a social choice function dominated by temporary dictator, controlled by a Congress of representatives of the people, can lead a country (rich in natural resources and inhabited by a people that accepts a set of values very respectful of the rules of civil coexistence) to a position of planetary almost-unchallenged domination.

In alternative, the temporary-constitutional monarch-French-President appoints the prime minister, presides over the Council of Ministers when he deems it appropriate, is ultimately responsible for the foreign and military policy of his country, but the government, which he has nominated, must obtain the confidence of the Parliament, which can force the prime minister to resign. The French system, conceived by De Gaulle at a time of a very serious institutional crisis, shows that even a country that respects the opinions of minorities can reform its institutions in order to make them more efficient. If perfect democracy (in the sense of Arrow) does not exist, with appropriate mechanisms of rotation and separation of powers that regulate the attribution of the different possible roles of a dictator, one can build a political system capable of favouring the political, economic and cultural development of a nation.

In the section of this essay dedicated to Montesquieu, the reader will find the tentative solution of the problem of finding the mathematical basis for the formulation of constitutions. It will be seen, while discussing Montesquieu thought, how we hope that the impossibility result by Arrow can be overcome.

Catherine the Great and Genghis Khan: Power Has No Gender

To understand the nature of power, to produce a taxonomy of the possible types of powerful men and women, to establish whether the exercise of power is truly inevitable, philosophers and scientists have written rivers of words and sometimes started heated debates. The understanding made possible by Arrow's theorem allows us to clarify many of the past controversies. However, this theorem has also opened many new and interesting questions. This circumstance is neither new nor specific to Arrow's theorem. Actually, this is one of the most remarkable features of the most important scientific results: in resolving some epoch-making issues, they invariably open up very broad new horizons. So vast horizons as to be able to say, and this may seem like a paradox, that thanks to these revolutionary results the number of unresolved problems is, at the end, enormously increased.

In fact, an important and old question still remains substantially open.

Does the gender of the dictator influence the way in which she/he exercises power?

In other words, the question is: Do males and females exercise power in different ways?

Our opinion is that there is total gender equality in all human activities, including the exercise of power. Since it is not unanimously accepted, we believe that our opinion must be argued for.

It is clear that there are insurmountable biological limits (and one cannot be accused of male chauvinism if one notices this circumstance) between genders: for example, if it is certain that Genghis Khan has managed to have at least 20,000 children and grandchildren, obviously this performance is simply impossible for a woman. However, perhaps we are wrong to give too much importance to this observation: after all, the mother of Genghis Khan had at least 20,000 grandchildren and great-grandchildren, and therefore, in the natural history of humanity, there are both males and females to have a very large offspring.

© Springer Nature Singapore Pte Ltd. 2019
F. dell'Isola, *Big-(Wo)men, Tyrants, Chiefs, Dictators, Emperors and Presidents*,
https://doi.org/10.1007/978-981-13-9479-9_3

There are very rigorous mathematical theories that study the strategies of transmission of the genetic genotype as implemented by individuals of each living species. These theories have made us understand a lot about the ethology of the species, and in particular, of the primates, apes and human beings. We believe that the union of these theories with game theory, dynamical systems theory and decision theory can still be very fruitful.

Many phenomena remain to be understood. For example, the origin of a specific behaviour observed both among humans and among bonobos: For what evolutionary reason do dominant females kidnap and raise the children of females submitted to them in the pecking order? In short: What is really the evolutionary mechanism that triggers this behaviour? It should be noted that this behaviour is in common, at least, with human and apes ethology.

We have not been able to find anything in the literature to answer these questions (and many others). We will try to formulate a conjecture on the lastly formulated questions later in this essay.

Genghis Khan

In the history books, Genghis Khan is best remembered for his exceptional talents as a unifier of his people, a man of government and a capable leader. The Mongol Empire (at its height in the thirteenth century) managed to dominate a large part of Asia and a big portion of Europe, too. A direct witness of his qualities as a statesman was Marco Polo: we know from reading *The Million* that Genghis Khan was certainly a skilled administrator of his Empire, careful in managing his finances, tireless in caring for public affairs.

One can reasonably ask: Why has a man devoted all his energies to the welfare of his people? What force can lead to such an extreme sacrifice of self?

The answer to this question, which, if one wants, could also be discovered in the folds of the chronicles of the time, has recently been found thanks to the modern techniques of genetic engineering. If all kings and emperors (and even popes, patriarchs and bishops) accepted, in a metaphorical sense, the title of "father of their peoples", Genghis Khan did so much because he aspired to become (truly!) the "genetic father" of the greatest possible part of his people. In fact, and in a few years, he managed to spread his DNA very widely. So widely that the Y chromosome of Genghis Khan has remained indelible in the Y chromosome of one male for every 200 human males living today.

As many readers will already know, the Y chromosome is inherited, without modification, from father to son. Only spontaneous mutations change it with the passing of generations: the markers of Genghis Khan, after only eight hundred years, remain perfectly distinguishable in 16 million human males (almost all of Asian ancestry, of course).

The chronicles of the time, sometimes discreetly but always very clearly, indicate which was the reason that led Genghis Khan to so much sacrifice of himself: he was a relentless, tireless and efficient sexual predator. His wars of conquest always ended with the kidnapping of the most beautiful women in all subjugated towns, villages and territories. Special agents in his service selected the most beautiful virgins and took them to their lord. And, let us repeat ourselves, a very reliable Persian source from the thirteenth century attributes at least 20,000 children and grandchildren to Genghis Khan. Since there is a 50% chance of having a male child, it must also be deduced that at least 16 million women, nowadays, are also his descendants.

There is an important and further clue that indicates that in the DNA of Genghis Khan, there was the gene of the serial and compulsive sexual predator. All his children acted as he did. His nephew Kubilai Khan, in bringing the Mongolian Empire to its maximum expansion, despite already having 22 wives, seems to have added at least 30 virgins to his harem every year, coming at the rate of nearly three new concubines per month. Perhaps the life of Genghis Khan was risky and tiring, but he is certainly the human being with the greatest living descendants of all history today, considering that the "poor" Aztec Emperors, who had thousands of concubines, have seen their descendants mowed down by the diseases imported by (and also by the massacres perpetrated by) the Spaniard *Conquistadores*.

Catherine the Great, Empress and Autocrat of All the Russias

At this point someone could express harsh judgements on the male gender, beginning to remember the brutality of the genocides perpetrated by Genghis Khan, the violence of his military campaigns, his lack of scruples in getting rid of enemies and adversaries. The conclusion often reached is that the violence of males is aimed at domination, domination that has its worst forms in the overwhelming of females and in the unscrupulous exploitation of their bodies.

However, before arriving at too optimistic conclusions about human nature, as declined by the female gender, the reader should consider, in parallel, the figure of Catherine the Great.

Catherine II is commonly described as an enlightened Empress, and at first glance her ability to lead her people shows no flaws. But, let us carefully examine the way in which Catherine came to the throne. Of Prussian origin (she was born in Szczecin in 1729), by a German general and a noblewoman of an ancient family, she is married, just over sixteen, to the heir to the throne of the Russian Empire, Grand Duke Peter, future Peter III, Tsar of Russia. Who chooses her as the wife of Peter III? Empress Elizabeth! She wanted to designate her nephew Peter as heir to her throne and, for that reason, she felt entitled to choose his wife. Just as many fathers have done in the history of mankind, and always claiming to work for the best of their daughters.

Empress Elizabeth, with a behaviour that has also been observed in apes (see the story of Kanzi also told later in this work), took the two sons Paolo and Anna away from their mother as soon as they were born, to raise them in her private apartments and preventing a normal emotional relationship from being created between mother and sons. As often happens in the societies of bonobos, the dominant female, even in the palace of the Tsars, decided the fate of the last born children of the family. Moreover, as in all matriarchal societies and bonobos societies, the fatherhood of Catherine's children is extremely uncertain. It is not clear who Paul's father was: even if it cannot be excluded that he was really the son of Peter III, it is very likely that the real father was one of Catherine's favourite lovers, Sergej Saltykov. On the other hand, it is almost certain that Anna's father was Stanislaus Poniatowski, the future Polish sovereign, for whom Catherine literally lost her head. In history one can, therefore, find examples of men who have made a career thanks to their mistresses! In 1762, Catherine gave birth, in the utmost secrecy, to her son Aleksej. Most likely the father was Grigory Orlov.

Year 1762 is a crucial one in Catherine's life. During that year, her husband ascended the throne as Peter III of Russia; Peter, with his first political choices, alienated all the major power groups in the court, groups with which Catherine, instead, had established excellent relations; Grigory Orlov led a conspiracy to crown her, carrying out a coup d'état; Catherine dethroned her husband by locking him up in prison and not opposing to his assassination.

Since the history books are full of chronicles of the Empress' successes, it is not worth mentioning here a brief summary of the splendour of her reign. We must ask ourselves, however, why Catherine wanted so strongly to govern, why she always tried to choose the best collaborators to carry out her policies and with what criteria she selected them.

Probably a very evocative and certainly explanatory image of Catherine's personality is given by a statue that she herself would have appreciated very much: the Monument dedicated to her in the great Ostrovski Square in St. Petersburg. The figure of the Empress dominates from above, imposing and represented on a larger scale, while a series of figures, represented on smaller scales, are placed at her feet: these figures represent her most influential and important collaborators. Among them, in addition to the already mentioned Orlov, also Potemkin and Suvorov stand out, who are said to have been also among her most long-lasting lovers. In fact, it is accepted by modern historiography that Catherine had literally a multitude of lovers and many chronicles, also among those written during her reign, consider her, probably with a background of truth, a tireless sexual predator. In fact, she regularly got tired of her last "friend-in-bed". To get rid of him, she usually gave him a lot of money, granted him positions and privileges (e.g. Grigory Potemkin was created by her as Prince of Tauride) and allowed him to manage ambitious and lucrative enterprises.

Indeed, Catherine was a powerful woman, and as such had no time to lose. In order to satisfy her needs as effectively as possible, she used at least two of her trusted company ladies as «Eprouveuses». That is, as experts who test the quality of the product before presenting it to their sovereign. In short, a sort of «tasters of men», who selected the men to be labelled by the certificate of quality sufficient to deliver them to the royal men-eater: Catherine. The first taster in office was Countess Praskovya Bruce, wife of the Governor of St. Petersburg and the only person Catherine openly defined as her right arm, ever. When Catherine caught Bruce in the act of tasting the royal lover in charge, Ivan Rimsky-Korsakov, without having explicitly had the assignment of this task, she was furious with jealousy and decided to exile the two lovers and to choose a new right arm: the Countess Stepanovna Protasova. The author of the delation that led to the discovery of this forbidden love was Potemkin, worried about the influence that Bruce and Rimsky-Korsakov were gaining on the Empress. It seems that from that moment on, the lovers of Catherine were selected by means of the following process: Potemkin approved their political, human and reliability qualities, a doctor ensured that they did not have venereal diseases, Protasova ascertained their physical qualities, resistance and experience, and finally the Empress acquired the imperial right of their exclusive exploitation that lasted until she got tired of it.

The Empress also proved to be greedy with food: her slender profile of portraits as a young woman weighs down in those as an adult. Her gaze, however, is always deep and faithful to her soul: clever, intelligent, sharp and sometimes benevolent and full of humour.

Historians were not concerned too much for the consequences, in the politics of the time, and in the development of historical events, of Catherine the Great's strong sexual impulses and almost insatiable sexual appetite. Exactly as done for Genghis Khan, they focused on the effects of the rule of these sovereigns on the peoples of their empires. For Catherine, they considered the impact of her choices on the development of the sciences and the arts in Russia and in the world. Also, they focused on the influence she had on the history of Europe, by choosing to involve more and more Russia in the international competition arena. They, instead, neglected to consider her inner motivations. The calculation of the number of those of her lovers who managed to resist for a reasonable period of time, that is to say, enough time to have had an appreciable influence on their sovereign's choices, leads to a total number of at least 22 men. This is not a number that is more surprising than that: it could put the Empress men-replacement rate in a reasonable average for a free woman who lived in one of the many eras and societies in which the sexuality of women has not been violently repressed. However, Catherine did not live in one of these happy epochs and societies: only by using her imperial power did she have the possibility of ignoring the canons imposed on the women of her world and was able to behave exactly like a man of her time.

Acquiring with certainty historical truths about such delicate issues is not easy. Even after many years, the figure of Catherine the Great was, and still is, used for contingent political purposes. Catherine is considered a mother of the great Russian homeland and therefore her sexual habits have been hidden by almost all Russian governments, as they needed to trace back to hers the roots of their own political choices: as, for example, to justify the umpteenth occupation of the Crimea. However, the existence of salons in her private apartments in the Imperial Palace, which were exclusively dedicated for her erotic leisure activities, seems certain. The servants had limited access to these apartments: to the point that a system of pulleys could raise a laid table, from the lower floor to that of the erotic halls. In this way, Catherine could indulge freely and without the hindrance of the presence of the servants to both his favourite excesses. Many photographs of the objects and furniture that filled these rooms are still available. Probably inspired by the frescoes and objects, which had been just found in Pompeii and which today fill the room of the secret cabinet of the Archaeological Museum of Naples, these unique objects were designed to satisfy the sexual fantasies of the Empress. Actually, of all the furniture of these rooms, only some rare photographs remain public and the suspicion remains that Soviet bigotry led the conservatories of the Hermitage to destroy it to preserve the reputation of the progressive Empress. However, the testimonies collected for the documentary by Peter Woditsch «The lost secret of Catherine the Great» (Sophimage, ART, RTBF) seem to leave no doubt. Catherine probably dragged her favourite lovers into her private rooms and dominated them there. Without going into details that are ultimately useless, consider, for example, the riding chair, whose existence and use are certain. The decorations of the armchair are so explicit and didactic that no unfortunate man, after having seen it, could ever claim not to have understood which was the performance that the Empress was asking to him.

Some Initial Conclusions About the Relationship Between Power and Gender

In the summer of 2018, we have seen the explosion of a scandal, which involved a famous director-actress. While she was accusing a male director of having asked her for sexual practices in exchange for help with her career, her current partner paid a large sum to a younger actor who accused her of rape (for having sex with him before he reached the age of consent, as required by the country in which the events happened). In 2019, a woman teacher 36 years old has been arrested for having had sex with a 13-year-old boy, who was her student. She delivered a child whose father was such a young boy.

Without wishing to enter into the specific facts and without wanting to pronounce on their possible justification, it remains clear that no difference can really be determined between male and female behaviour, not even in this kind of sexual behaviour and also when taking into account the differences between male and female physiology and physical force.

It does not seem to us that the theoretical effort needed to clarify the many obscure aspects of the dictator's behaviour (and, in particular, his management of power) should take into account the gender of those who exercise power. Obviously if the physical force must be used in the struggle for power, or in the exercise of power, then it must certainly be taken into account that a male gorilla weighs at least three times more than a female gorilla and that a man is generically stronger than a woman. We must, however, take into account that for this difference of strength and tonnage, the females of the bonobo have compensated in three ways: systematically selecting as fathers of their sons some males decidedly thin; allying themselves in bands of females that react in a coordinated and collective way to the aggressions of each male; educating their sons to be obedient and to follow their mothers all the time in the search of food. Actually, the male ethologists, at first, misunderstood the bonobos' ethology because of a macho prejudice. Indeed, they observed the existence of nuclear couples of bonobos, one male and one female, living all the time together and sharing their life: they wrongly concluded that bonobos were given a wonderful example of male/female family based on a strong monogamic link. Nobody among these first observers did suspect the truth, i.e., that they were observing mothers faithfully followed by their sons. Only after having more carefully studied the bonobos social life very carefully, other scientists understood fully the social role, which is played in it by sex.

As we can see, those women, who have allied themselves against the male sexual predator, have not invented anything really new: they have only applied a strategy that has already been widely exploited by bonobos.

Some Open Problems For A General Mathematical Theory Of Social Structures

Arrow's theorem, based on Condorcet's ingenious intuition, is only a first important step towards a **General Mathematical Theory of Social Structures.** Yet, the understanding that has been made possible by this fundamental step forward has had a great impact on our understanding of a myriad of social and ethological phenomena. The present work has the ambition to bring to a wider audience not only the fundamental ideas of the theorem but also to describe its most important implications in applications.

However, we are far from being able to control social phenomena with the precision with which we control certain physical phenomena, such as those described, for example, by electromagnetism, quantum mechanics, structural mechanics and thermodynamics. New mathematical instruments and new models must be developed. Some of them will be completely original while others will appear as simple adaptations and developments of methods and theories already known. Probably among these last tools, we have to include the Calculus of Variations, the Theory of Optimisation and the Lagrangian Dynamics.

Once a sufficiently predictive theory will be elaborated, it will become possible to try, for example, to establish the optimal constitutional mechanisms that must regulate the modalities of selection of a dictator, without risking that democracy degenerates into a tyranny. A preliminary problem to be solved, to this aim, could be the following:

In which situations does a democracy need, if it wants to survive, the election of a dictator?

A society, which can select the right dictator to face a specific social crisis, has a better chance of survival. However, it must be able to get rid of the dictator when he tries to become a tyrant (we use here these two words with the meanings which are explained in the first appendix at the end of the book). This is a very delicate issue, which will require the development of complex mathematical theories generalising those used for formulating Arrow's theorem. These theories will have to enable us to understand,

© Springer Nature Singapore Pte Ltd. 2019
F. dell'Isola, *Big-(Wo)men, Tyrants, Chiefs, Dictators, Emperors and Presidents*,
https://doi.org/10.1007/978-981-13-9479-9_4

for instance, the differences between France at the end of the Fourth Republic and the Italy of the Notables (beginning of the twenties of the twentieth century) or Germany of the Weimar Republic. In fact, the impasse in French democracy was resolved by a democratic dictator (De Gaulle) while the evolution of Italy and Germany, between the two world wars, led to two of the worst tyrannies in human history.

We believe that it is reasonable to accept, once again, how truly plausible the visions that were presented by Asimov (see, e.g., his Foundation Trilogy) in the form of science fiction stories. Asimov did imagine that a talented mathematician, in a very distant future, could develop the «psycho-history» and that this theory will be able to predict, for example, in what situations societies evolve from the stage of "parliamentary democracy blocked by minority vetoes" towards a «democratic dictatorship» of the Gaullist type, and to establish, instead, when their natural evolution leads them towards a «tyranny».

A Casual List of Problems, Which Could Be Treated by the Theory of Social Structures

Many other important problems need to be studied by the mathematicians interested in applications to social sciences. Let us consider, for instance, those problems, which concern the dynamics of societies immediately after a revolution.

Why did both the French and Russian Revolutions lead to a period of «terror»? Why was Robespierre quickly removed from power while Stalin remained firmly in the saddle for many years?

Why all the Italian revolutions did not lead to the development of new regimes? Why the necessary institutional reforms in Italian States were initiated only when the French Revolution was imported with the armies of Napoleon?

Are there societies that are more subject to the affirmation of a tyranny? Why does Russia seem incapable of getting rid of the Tsars?

Is it possible to strictly characterise those properties of a social structure that determine its tendency to develop revolutions? And why do some revolutions fail at birth, others succumb to counter-revolutions while others lead to completely new social structures?

What is the difference between freedom and anarchy, between «just rule» and tyranny, between well organised society and repressed society?

Is it possible to understand the role of the educational system, teachers and politicians in the structuring of happily democratic societies?

Plato's Republic: A Pristine Textbook in Social Sciences

Pourquoi donc, (…) citez-vous un certain Aristote en grec? — C'est, répliqua le Sirien, qu'il faut bien citer ce qu'on ne comprend point du tout dans la langue qu'on entend le moins.

Why ever, then, (..) are you citing a certain Aristotle in Greek? — The reason is, replied the Syrian, that one must, actually, cite what he does not understand at all in the language which he understands the least.

Micromégas (1752), Voltaire, éd. Hachette, coll. «Classiques Hachette», 1991

All the problems mentioned in the previous pages have been formulated long time ago. They were debated at least since the fourth century BC. In the *Republic* by Plato it is reported the state of the art in the theory of constitution as understood in that epoch. We have found really topical, even nowadays, the ideas and concepts discussed by Plato.

In particular we have found really lucid and effective the description, given by Plato, of the populist drift, which too often is observed in democracies.

Taken from "La Repubblica" by Plato Book VIII from line 964 to line 1040

Our translation is obtained elaborating on the translation from Greek into Italian by Francesco Gabrieli

«When a democratic city, thirsting for freedom, comes to have as its leaders some bad cup-bearers and intoxicates itself with that freedom beyond the need, it happens that:

– If magistrates are not completely submissive and do not bestow freedom in very large amount, should be punished for being accused of being wicked and oligarchic;

– Those who obey the authorities should be victimised as voluntary valueless slaves, while those rulers who are similar to the governed should be praised and honoured in public and in private;

– That the father accustoms himself to be like the son and to fear his children, and that the son accustoms himself to be like the father, and to no longer be afraid or respectful of his parents, believing that this is the only way to be free;

– The teacher fears and caresses the pupils, and the pupils flout the teachers, and, in general, the young people put themselves on an equal footing with the old people, contending with them in words and deeds, and the old, lowering themselves to the level of the young, are filled with playfulness and pleasantness, imitating the young so as not to appear to be authoritarian or boring;

– Nobody should give any thought to those written or unwritten laws that regulate social life, because they believe that those who respect the laws are slaves.

In this excess of freedom, a bad plant takes root and develops: tyranny.

In fact, the excess of freedom is always converted into the excess of servitude, for the individual and for the state. Tyranny, a disease that ruined the oligarchy, becomes a disease that afflicts democracy in an even more virulent way, when it [democracy] falls into the excess of freedom. In fact, it is pushing something towards an excess whatsoever that produces, conversely, a great change in the opposite direction: in plants, in bodies and, not least, in political regimes.»

A Quest of Phenomenological Basis for Future Theoretical Developments

Since, in specific circumstances, a dictator is logically necessary, an efficient democratic system must provide both effective mechanisms for his designation and equally effective procedures for initiating and completing the process of his dismissal, when his task is over or if a drift towards tyranny manifests itself. Such control mechanisms should be sufficiently stable and effective to resist the action of people such as Robespierre, Stalin, Hitler or Mussolini, since, unfortunately, not all dictators are so wise as to be able to behave like Silla, De Gaulle or Garibaldi.

The General Mathematical Theory of Social Structures does not yet exist. The phenomenology of the dictator, that is the set of experimental observations available on the behaviour of dictators, is instead enormous: it fills the books of history.

The competitive advantage obtained by a group when his dictator is altruistic and clever is enormous. Consider, for instance, the successful appearance on European politics of Russia, when Peter the Great became its autocrat. Many famous statesman or politician did get fame and glory because of their abnegation in serving their people. Many others became infamous because of their incapacity or arrogance or violence. However, when judging about the action of a known, be its fame good or bad, dictator one should be careful. In fact, historians are influenced by the propaganda of the parties that eventually won the political struggle and some wise and capable dictators may have a completely undeserved negative or positive fame.

For instance, it is clear that Suetonius, while writing, was serving the party of the Senate. As the Senate had opposed Nero and Caligula, then Suetonius describes each of them as insane and crazy dictators, who burnt Rome or nominated a horse as senator. One could argue that this second fact, which was a clear counter-propagandistic action, could have even improved the quality of a probably corrupted Senate, and that, as proven by a careful historical investigation, Nero did his best to improve the conditions of his empire, even by violently opposing the corrupted aristocracy which was undermining it.

What we want to underline here is that nobody remembers the dull, incapable, deceitful, dishonest and corrupted dictators, if they managed to moderate their negative actions so that they are not remembered as champions of evil. There were mediocre dictators, who did not manage to do very much for their people and simply survived enjoying their privileges without really influencing the course of history. Indeed, the great majority of dictators do not influence either positively or negatively the events occurring to their people.

On the other hand, the great majority of capable individuals do not manage to reach power positions, so that they are frustrated in their efforts to direct their people towards glorious destinies. Actually, very often potentially capable dictators do not have the talents needed to reach power and are defeated in the competition. The winners in this competition usually are individuals who, while having the talents to prevail in the dictatorship selection process and to actually become dictators, are instead unfit to be effective leaders.

We are sure that history is full of these unfortunate failures and these intelligence wastes: however, only a small percentage of the possible positive opportunities lost for human societies are actually documented. In the next pages, we will discuss some examples of such lost possibilities. We can really talk about the impotence of intelligence, as blocked by joint action of stupidity, arrogance, ignorance, avidity, lust and blind hunger of money and personal privileges.

The general theory of societies should develop methods of constitutional engineering, which favour the ascension to dictatorship of those individuals whose true interest is the vainglory which pushes them to the desire to become famous because they could lead their people towards happiness and success.

Only a few specific figures of leaders and dictators have been described in the next pages. In them, a brief description of their behaviour and nature has been attempted. Inevitably, the phenomenology described is partial, as the experimental evidence one needs to account for is really enormous. Probably the presented phenomenology is not even representative of the limited range of social mechanisms, which we wanted to present. However, we believe that it is indeed sufficient to give an idea of the importance of the challenges that modern mathematics must accept and overcome.

Phenomenology of Leadership

Even if we are aware of the fact that it may appear as a completely arbitrary differenti-ation, however, we will try to distinguish, in the following pages, between leadership and dictatorship, giving to the first word a slightly more positive meaning. On the other hand, as Machiavelli teaches us, abstract concepts of ideal morality can be deadly for a social group, if their leaders, dazzled by utopias, forget to face the most brutal aspects of reality. It is therefore doubtful that "moral" leaders be better than resolute dictators, for their own people.

It is remarkable that exactly those political systems, like the British Empire, which officially rejected the political thought of Machiavelli have been, in practise, the most Machiavellian ones. Actually, to be successful, they did not have alternatives.

On the contrary, many leaders were obliged to experience the impotence of intel-ligence against the conjoint action of many short-sighted opponents, guided only by their immediate interests. The presumed intellectual superiority of these leaders was frustrated by a diffused cynic ignorance. Ultimately, we can say that the leaders we are talking about have often reached the highest peaks of arrogance, in particular when they have ignored the true force of their competitors, simply assuming that not so clever opponents were bound to succumb. If one believes that the simple fact of being right will automatically imply that other people will accept to follow one's indications, then he suffers from a pure lack of the sense of reality.

A really outstanding leader manages to guide his group towards the right direc-tion NOTWITHSTANDING the fact that nearly all group members do not man-age to discern the actions, which are truly convenient for them.

Before continuing the announced phenomenological analysis of leadership, the reader will allow us two short digressions, which are only apparently out of the subject established for this discussion.

© Springer Nature Singapore Pte Ltd. 2019
F. dell'Isola, *Big-(Wo)men, Tyrants, Chiefs, Dictators, Emperors and Presidents*,
https://doi.org/10.1007/978-981-13-9479-9_5

Leaderships and Unifications: Analysis of an Historical Necessity

In many scientific and academic debates, the author was «accused» to be Neapolitan. Notwithstanding this circumstance, he will not be afraid of discussing the contribution given by Neapolitan culture to science. However, it is not the need for affirming a cultural heritage, which motivates some of the following pages of this essay. Instead, the true motivation of the presented analysis is more general (and really topical!) in our epoch.

We aim here to describe the process of unification of Italian peninsula under a unique State.

This process deserves the attention of the scientist because of its intrinsic interest: humankind can optimise its chances of survival only if it will manage to behave as a whole, unique entity, capable to face the challenges, which our universe presents continuously.

In particular, for European citizens, we believe, the unification process experienced by Italy deserves a particular attention, at the beginning of XXI century. Europe, for preserving its culture, its economical wealth and its influence on the future of humankind MUST unite. However, economical and political Unions are not simple processes. Indeed in Italy, nearly one third of the country, as a consequence of the unification process, was inflicted serious social and economical damage. Hopefully, Europe wants to avoid repeating the mistakes made in the past, when, with violence and corruption, a State was forced to dissolve without any serious warranty of the respect of the rights of its citizens, of their traditions, their language and their culture.

In absence of a well-established theory, which we claim is urgently needed and whose construction hopefully will attract the intellectual forces of many mathematicians, the only possible way for designing a unification process consists in the careful empirical study of phenomenological historical evidence.

Giambattista Vico pioneered this kind of studies, but already Cicero was aware of their importance. In his work, Cicero states that «**Historia magistra vitae**», that is: «**History [is] the teacher of life**».

For being clearer, it is worthwhile to refer here the complete sentence that is:

Historia vera testis temporum, lux veritatis, vita memoriae, magistra vitae, nuntia vetustatis (Cicero, De Oratore, II, 9, 36),

which we can translate as follows:
«**History is true witness of the times, light of truth, life of memory, teacher of life, messenger of antiquity.**».

In fact, we believe that the outstanding importance of history must be sought in its more fundamental role in the advancement of human science.

History has to be our teacher.

In this sense, we feel to be disciples of Giambattista Vico. We will not talk about Vico in this essay, as his figure alone would deserve a treatise: he was one of the greatest (and relatively least valued) among Neapolitan intellectuals. He was the greatest, more generally, among all those philosophers who, in Western civilization and his epoch, tried to transform the study of history into a science.

Southern Versus Northern Leaders

Each country is the North of another country, but also the South of some other country.

This sentence was what a wise old Neapolitan grandmother said to give good advice to her grandson, while he boasted about the civilisation he believed he had observed in Northern Italy.

Attention, however when dealing with the concept of North (if you want to understand it as synonymous with the «advanced part of a Nation») we must be very careful: during the Lombard domination in Italy, a Langobardia minor was located in what is now the South of Italy (Duchy of Benevento) and the Byzantine thema (i.e. an administrative region of the Eastern Roman Empire) of Langobardia included the areas of Southern Italy not occupied by the Lombards. All these territories were seen as «North» by the inhabitants of North Africa.

Moreover, in France (but also in Germany or England), the North is to the South! Explaining the wordplay: the most developed part of the Nation is in its Southern part, while the North of France is considered in France (in many ways and from several points of view) as, in Italy, people considers the South. To be convinced of this situation, it should be noted, for example, that the movie «Welcome to the South (Benvenuti al Sud)» is a remake, based on an ethnic adaptation, valid for Italy, of the French movie «Welcome to the Sticks (Bienvenue chez les Ch'tis)» . In France, it is customary to say that: «to the North one arrives crying (*for the desperation of having to go there*) and leaves crying (*for the displeasure of having to go away*)». In Naples, the same thing is said but replacing North with Naples.

Since in Italy, and in the present historical moment, North indicates «the most developed and best organised part of the country» and since this essay is written by a South Italian during 2018, the term North will be used with this meaning: however, it is not so far away the time when the South of Italy was far more advanced, than its North. It is enough to think about the Kingdom of Two Sicilies before Italian reunification. A 2010 study by the Bank of Italy (see Ciccarelli, Fenoaltea 2012) reconstructs the industrial trend at the level of individual provinces, for the years from 1871 onwards. The study concludes that:

"At the provincial level, the link between industry and economic success in general is weaker than the previous literature assumed. Further disaggregation reinforces the main revisionist hypothesis suggested by regional estimates. Provincial indicators confirm that a decade after the Unity the ancient political capitals remained the centre of (artisan) manufacturing, that the under-industrialised areas were the Adriatic and Ionian peripheries of the largest entities, and that the industrial backwardness of the South, evident on the eve of the Great War, was not the legacy of Italy's pre-unification history".

Naples is the northernmost city of the kingdom of which it was the capital and Frederick II Hohenstaufen founded his University there for this very peculiarity. On the other hand, Sicily has always considered with impatience the dominion, which the technocratic Neapolitan aristocracy tried to exercise on the island. Naples had one of the oldest universities in the world, perhaps the first state university ever: its intellectuals were perfectly able to understand what it was necessary to do to bring progress, independence and prosperity to their State. In their own way, the Neapolitan technocrats tried to bring the whole of the Italian South into Europe, during the First Industrial Revolution. However, they failed: these pages will attempt, among other things, to give an explanation of this failure.

The question we are trying to answer is:

If the South is the part of Italy which has been the wealthiest, most fertile and most hospitable; was fairly well endowed with raw materials; has seen the birth and development of two of the oldest and most glorious universities in the world (in Naples and Catania); has produced revolutionary ideas in economics (think of the work of Antonio Genovesi or Ferdinando Galiani), still used for the government of the currency; has had scientists and philosophers of the greatest value (for example Giuseppe Mercalli, Renato Caccioppoli, Ettore Majorana); has forged European law with its jurists (and it is not possible to try to list here the most important among them, so many have been in the history of Italy),

how it is possible that this same part of Italy is now the poorest, so poor as to force its best children to emigrate, emigration which in the last hundred and fifty years has assumed «biblical» dimensions?

Indeed only in the last ten years, nearly two millions persons, mainly intellectuals or highly qualified workers, left South Italy to go to work abroad or in other Italian regions: consider that the whole South Italy has now less than 20 million inhabitants.

Too simplistic explanations, which refer to the innate ineptitude of the southerners, we will not even consider.

If anybody thinks that an argument based on our biological knowledge is necessary, we will remember that a vast group of geneticists, including Luigi Luca Cavalli Sforza, has demonstrated the total inconsistency of the very concept of genetic peculiarity of the traits of the social group that some call «southern». The evolutionary tree of the human genome and the history and geography of human genes clearly demonstrate that the answer to the problem posed is not to be found in the presumed endemic spread in Southern Italy of the gene of crime and the inability to comply with the laws. Even some politicians who have achieved great national fame by initially advocating such theories have now corrected the shot: they have sought even in a «deeper» South (particularly in Africa) the gene of violence and incivility. Unfortunately, these politicians refused to help «the southerners at their home», and interrupted—instead of making more incisive—the action of the «Cassa del Mezzogiorno» (which can be translated in English as «Fund for the South») just when it was about to be successful.

The critical Northern reader, before criticising the last statement, should consider that for making effective the reunification of West Germany with East Germany, the German reunified state spent, in the last 30 years, for each East German between 5 and 10 times more what has been spent for each South Italian between 1950 and 1992. This same reader should also consider that what has been spent by the Cassa del Mezzogiorno amounts to more or less one-half of the amount of economical resources «transferred» from South to North immediately after the reunification. Finally, we recall the highly qualified analysis by [Viesti 2013]. In it is proven that the large investments, which were planned for South Italy actually, in a great part, produced only profits for North Italian companies and had a small impact on the economy of South Italy: these investments instead of decreasing the North–South gap had the effect of increasing it.

The same kind of politicians now pretends to want to «help at their home» those desperate who, risking to die drowned, are transported, by criminal organisations, across the Mediterranean to Italy, after having started their emigration from the most underdeveloped countries of Africa.

In reality, the Neapolitan technocrats of the Kingdom of the Two Sicilies were the first to pose a «southern question», where the South was, in that context, Sicily and Calabria.

The administrative, economic and political division between Sicily and the continent has ancient roots: the Arab conquest of Sicily, the Norman reconquest, the Sicilian Vespers and consequent duplication of the Kingdoms in southern Italy: a King of Anjou family in Naples and an Aragonese King in Sicily. The southern question, even in the limited context of the economic and political history of the Kingdom of Naples, had deep roots and its understanding and the consequent solution was not easy.

The Neapolitan technocrats were northern enough to have, in general, a solid sense of the state but also southern enough to understand the subtle Arab–Byzantine mechanisms that have always regulated, for example, Sicilian society and therefore they considered themselves capable of effectively affecting these mechanisms to improve the standard of living in Calabria and Sicily.

Probably, it was for this reason that the Sicilians preferred to give themselves to the Piedmontese: they believed they could manage to leave their way of considering the state unchanged by formally giving power to men who could not even understand their way of being and living. The Leopards (do you remember the novel by Giuseppe Tomasi di Lampedusa and the movie by directed by Luchino Visconti and starring Burt Lancaster, Claudia Cardinale and Alain Delon?) wanted to apparently change everything so as not to change anything.

For the same reason, in history, there has been only a general and politician who achieved a real victory over the Sicilian refuse to accept any form of truly efficient State. This refuse has had various forms and came back several times, again and again. We refer the Neapolitan General about whom we will talk in what follows: Carlo Filangieri.

Leaders from South Aim at Unreachable Ideals?

Dans ses écrits un sage Italien dit que «Le mieux est l'ennemi du bien»
In his writings a wise Italian says that «The best is the enemy of the good».
[Alternative translations may be: «The perfect is the enemy of the good» or «The better is the enemy of the good.»]
Voltaire cites this aphorism in "La Bégueule" 1772 ("The prude woman" in Contes en Vers Œuvres complètes de Voltaire, Garnier, 1877, tome 10 (p. 50–56)). He also repeats it (without citing his sources) in Italian (Il meglio è l'inimico del bene [«l'inimico» is the ancient version of «il nemico», whose structure remained in French, for "[the] enemy"), in the article "Art Dramatique" ("Dramatic Art", 1770) included in his Dictionnaire philosophique.
We like to underline that the traditional Italian aphorism: «Il meglio è nemico del bene» is attested since 1603: see e.g. Proverbi italiani (Italian Proverbs), by Orlando Pescetti (p. 30, p. 45).

The best is the enemy of the good, seems to be the motto of many among the honest Italians, especially from the South. On the other hand, also dishonest Italians exploit it very often: in order to stop the actions of honest persons, the dishonest ones always remember them that: «maybe what you are doing is good: BUT one should do its BEST».

While the honest ones try to reach the best, the dishonest ones have the opportunity to attain their personal interests, which would have been made impossible by the attainment of the good.

We will come back to this motto and its interpretation at the end of this chapter.

Gaetano Filangieri (1753–1788), has been one of the most eminent philosophers of law in modern history. Gaetano's most original work was «The Science of Legislation», which his premature death left incomplete. Since the necessary rationalisation of legal thought was very clear in Gaetano's works, they were widely appreciated in Europe and USA. Among the many admirers of Gaetano's thought, the most famous were perhaps Napoleon and Benjamin Franklin. Napoleon, just to honour Gaetano's genius, took care to have Gaetano's son Carlo educated. The work of Gaetano Filangieri directly or indirectly contributed to strengthen the theoretical basis on which the modern concept of the constitution was developed and put into practice. In fact, already in 1755, **Pasquale Paoli** promulgated the first modern constitution, the one that had the ambition to transform Corsica into a free and independent state. Paoli was a student of **Antonio Genovesi**, professor in Naples, perhaps the first professor of Political Economics in the world: actually, Paoli formulated his unfortunate utopia in Naples. He wrote:

We are Corsicans by birth and feelings, but first of all we feel Italian by language, customs and traditions... And all Italians are brothers and must be united before history and before God... Like Corsican we do not want to be servants or "rebels" and as Italians we have the right to be treated the same as other Italians... Either we win with honour or we die with weapons in our hands... Our war of liberation is holy and just, as God's name is holy and just, and here, in our mountains, the sun of freedom will appear for Italy.

(Quoted in Carlo Silvano, Breve storia di Nizza, Brief History of Nice, Youcanprint, 15 July 2015, ISBN 978-88-911-9757-3. URL consulted on 21 March 2017.)

The utopia, to which Paoli dedicated his whole life, will dazzle many politicians who interacted with the intellectual milieu which developed around the University of Naples during the Enlightenment. Also, the unfortunate General Murat met his miserable fate by launching a proclamation that resounds with Paoli's dreams.

*Proclamation of **Joachim Murat** to the Italians. Proclamation Of The King Of Naples.*

Italians! The time has come that your high destinies must be fulfilled. Providence finally calls you to be an independent nation.[...] And to what extent do foreign peoples claim to take away from you this independence, the first right, and the first good of every people? In what capacity do they rule over your most beautiful lands? What is the extent to which your wealth is appropriated in order to transport it to regions where it was not produced? Under what title do your children finally are kidnapped to you and sent away, destined to serve, to languish, to die far from the graves of your ancestors? [...] No, no: clear Italic soil of any foreign domain! Once masters of the world, you expiated this perilous glory with twenty centuries of oppression and massacre. May it be your glory today to have no more masters. [...] Eighty thousand Italians from the States of Naples marched under the command of their king, and swore that they would not ask for rest until after the liberation of Italy.[...] All the national energy and in all forms must be displayed. It is a question of deciding whether Italy should be free, or whether it should bend its humiliated front to serve again for centuries. The fight is decisive: and we will see well assured for a long time the prosperity of a beautiful homeland, which, still torn and bloodied, excites the greed of many foreign nations. [...] Rimini, 30 March 1815. Gioacchino Napoleone

It is typical of the dictators of the South, to be animated by sincere and visionary reforming intentions!

They cultivate the illusion of being able to have recognised their own reasons by virtue only of the evidence of the facts and the strength of the arguments.

Joachim Murat, Carlo Filangieri and many others pretended to forget that history teaches us some simple facts. For instance that a deliberative body, when driven by populistic demagogues, can vote everything against evidence, even that a white horse is actually black, or that people will never pursue their own interest as a whole, as each individual will pursue his own particular interest deciding to ignore the well-being of the community to which they belong.

These deluded and idealistic dictators of the South have systematically seen their dreams clash against the joint action of the opportunistic dictators of the North, allied with the most eminent exponents of the other kind of dictatorship which is found in the South: the criminals and the opportunistic, falsely honest, pursuers of exclusively their own interest.

Coming back to the philosopher Gaetano Filangieri: he contributed to the inspiration of the American Constitution, also through his learned correspondence (in French) exchanged with Benjamin Franklin, and continued the elaboration of a very advanced philosophical thought, a genuine product of the great Neapolitan culture which also saw **Giambattista Vico** and **Pietro Giannone** as protagonists. A parenthesis must be opened for this last philosopher whose destiny is common to many southern intellectuals. Even if he did not manage to be burned alive like **Giordano Bruno**, Pietro Giannone can be considered a typical representative of the southern idealist intellectual. Apparently an upright servant of logic and a faithful servant of his princes, Giannone had no hesitation in criticising the Church, the Habsburgs and the rulers of the Serenissima (Venice Republic), and ended up dying imprisoned in Turin. Giannone was never able to fully pretend a compromise and always seemed to put his principles before his own life. Critics accused him of plagiarism and his controversial figure seems to reflect in one person the duplicity of the behaviour sometimes observed in southern intellectuals.

Instead, Gaetano Filangieri was not an intellectual prisoner of provincialism: open to the theories elaborated all around the world, he knew and elaborated the theories of French philosophers, in particular, Montesquieu and Rousseau. The Science of Legislation discusses and brings to light the social injustices of the time, which were common in the great European metropolises of his epoch: London, Naples of Bourbons, as well as the many other European capitals (like Paris, St. Petersburg, Madrid, Venezia, Amsterdam, etc.). They were all afflicted by the unbridled luxury of the feudal privileges of the aristocracy and the clergy: these two classes, together, exploited without many scruples the people.

Gaetano Filangieri understood that the problems of the societies of his time were related to the widely spread privileges of dominant classes and not on the lack of competence and good wills of monarchs.

He recognised in his contemporary times the replica of the contraposition between the "populares" (i.e. the political party which supported Julius Caesar) and "optimates" (i.e. Senatorial Aristocracy). In favour of lower classes, Gaetano asked the Crown to be the agent of a "peaceful revolution" and to transform itself into an enlightened monarchy. This transformation had to be achieved through a serious reforming action to be implemented «on the legal instruments». Gaetano Filangieri understood, theorised and developed part of the modern doctrine of «the rule of law», as a fundamental tool for the well-being of societies. In his analysis, it is mostly important the affirmation of the need to carry out a codification of laws, a fair distribution of economical resources and also a qualitative improvement of public education. As far as criminal law is concerned, Filangieri conceives the modern definition of crime, which is nowadays accepted, in the greatest part world legislation. In his own words: *"Not all actions contrary to the law are crimes, not all those who commit them are criminals. Action separated from will is not imputable; the will separated from action is not punishable. The crime consists therefore in the violation of the law accompanied by the will to violate it"*.

The fortune of the Filangieri's work was vast in Europe and in the whole world. Nearly obviously, because of his deeply rooted rationalism, this work has been included, by the Catholic Church in 1784, on the Index of Forbidden Books.

It is ironic that in Naples, where a renowned juridical school so greatly contributed to the modern formulation of the concept of Constitution, exactly in the Kingdom of which Naples was the Capital, it was never possible to establish, in a stable way, a modern Constitution.

Why this could happen?
Probably because in Naples the best has been always preferred to the good!

An Exemplary Idealistic Southerner: Carlo Filangieri

Credimi, per chi ha un po' d'onore e di sangue nelle vene, è una gran calamità nascere napoletano.

«Believe me, for those who have a bit of honour and blood in their veins, it is a great calamity to be born Neapolitan» (as used before the reunification of Italy here the adjective «Neapolitan» means «subject of the Kingdom of the Two Sicilies»).

(from a letter to his son, as cited by Elena Croce in La patria napoletana, The Neapolitan Fatherland).

Carlo Filangieri Prince of Satriano, Duke of Cardinale and Taormina, Baron of Davoli and Sansoste, Lieutenant-general of the royal domains beyond the Lighthouse, Prime Minister of the Kingdom of the Two Sicilies

Carlo Filangieri, was a Neapolitan soldier and statesman, son of Gaetano Filangieri, the celebrated philosopher and jurist, whose intellectual achievements we have just shortly described. Carlo, a politician believing in Enlightenment, completed his father intellectual achievements, trying to put them into practice.

There is a very meaningful episode of his life, showing his temper, and his idealist vision of life, which alone resumes the tragic destiny of impotence of his proud courage and dignified intelligence.

While being in Spain, serving Napoleon during his Spanish campaign, he was provoked by the arrogant Milanese Napoleonic General Franceschi. This man complained about the behaviour of a Neapolitan officer, insulting him, but also generalising his comments in a racist way. He claimed that all the Neapolitans are imbecile. The younger Carlo, being only Captain, dared to answer proudly to the General remarks and the dispute ended with a duel, challenge that was strictly forbidden and punished as a very serious crime. The duel ended with the killing of the General and Carlo was judged by Napoleon personally. The reason of this privilege were two: first of all, Filangieri was a famous name in Napoleonic epoch, as Gaetano was considered to have been the most outstanding jurist in Europe, and second because Carlo had become a protégé of Napoleon for his outstanding merits.

When Napoleon heard the story, and understood the motivation, which had animated Carlo, he decided to find a punishment comparable to the death penalty which was inflicted in these cases.

Napoleon decided that such a punishment was to allow Carlo to try to prove that he was right in believing that Naples could become one of the most advanced countries in the world. He ordered Carlo to serve as an officer the Neapolitan Napoleonic State: as the flow of events will prove, this was a terrible punishment indeed!

The Main Events in the Life of Carlo Filangieri

Carlo was born 10 May 1784 at Cava de' Tirreni, near Salerno. Exploiting the turmoil consequent of Napoleonic campaigns, at the age of fifteen, he started a travel pretending to have the intention to reach Spain for completing his studies. Instead, Carlo, who was willing to undertake a military career, diverted towards Paris, as he had obtained a recommendation introducing him to Bonaparte, who was then "the first consul". However, more effectively than the letter that he was carrying, what protected him, and positively influenced Napoleon, was the fact that he was the orphan of Gaetano, whose juridical works the well-cultivated first consul had carefully studied.

This episode, probably, made Carlo believe that the value of merit is always recognised: this is the delusion which he will see systematically frustrated in the rest of his life. Very few politicians systematically preferred to support meritorious men (or their children), during human history. Napoleon was one of these few!

Carlo was immediately sent to the Military Academy at Paris, Prytanée national militaire, the best French military school, getting in two years the grades of officer, being the first out of 300 students. Napoleon had been right in supporting such a promising young man.

In the subsequent years, he could prove how brave, clever and faithful to the Napoleonic cause he was. Indeed, he participated in virtually all Napoleon campaigns, fighting with distinction, and being wounded several times and repeatedly promoted. He started with the expedition in the Low Countries; then he was at Ulm, Maria Zell, Austerlitz, Captain under Masséna in Naples and subsequently in Spain, where, as we have already seen, he killed General François Franceschi-Losio. Sent back to Naples, in what will become a sort of Dantesque retaliation punishment, he became General and served under Joachim Murat fighting against the Anglo-Sicilian alliance. After the fall of Napoleon, he followed Murat's effort for the reunification of Italy, against Austria, and fought heroically during the battle of the Panaro (1815), where he was seriously injured. In the Panaro battle, Carlo nearly sacrificed his life for helping his King by attacking a bridge with few soldiers, after having understood that the General in charge of this action refused to obey the order.

After the Bourbon restoration, King Ferdinando initially confirmed Filangieri in his rank and command. However, during the rebellions in 1820, Carlo adhered to the

Constitutionalist party and followed General Pepe who tried to resist the Austrian army, called by Ferdinando to re-establish his autocracy. As a consequence of the defeat against the Austrian, he was dismissed from the service. For the first time in his life, he had to pay the coherence to his ideals with a loss of positions, influence and social status: indeed Carlo's family although being very ancient and noble was not rich.

In this circumstance, the fate favoured him a second time, as it had done a first time when allowing his meeting with Napoleon. Suddenly some other possibilities to try to apply his intelligence at the service of his State were opened by an unexpected event. The eccentric sister of his father, who had married the extremely rich and noble prince of Satriano, Filippo Fieschi Ravaschieri, decided to nominate Carlo as her heir, with the explicit intent to support his intelligence and political action. Carlo became himself Prince of Satriano, inherited an immense wealth and numerous, ancient and rich fiefs. Having been dismissed from every public service, he retired in his newly inherited possessions in Calabria, hoping to be able to play a role in the economical, social and technological development of this unlucky Italian region. This dream, also, was frustrated by the insurmountable wall constituted by bad habits, ancient privileges, miserable local interests and economical unfair competition of foreigner powers.

The successor of Francesco I, Ferdinando II, tried to start a season of reforms in his kingdom.

At the beginning of the troubles of 1848, Filangieri became counsellor of the king, being recognised as one of the most competent and capable statesman in his kingdom and because of his great economical and political influence in the affairs of both Calabria and Sicily. Actually his wife, also, belonged to a very influential and rich Sicilian family, and Carlo could be of great help in support of the reformative policies of Ferdinando II. Once consulted, based on his experience and studies, he proposed to the king to finally concede a constitution. Notwithstanding that the king had followed the advice, in February 1848, the Sicilians, again incited and supported by Britain, formally announced secession from the Neapolitan kingdom.

Recalling the military skills that Filangieri had shown in his life, the king gave to him the command of an armed force to re-conquer the control of the Sicily. On September 1848, he landed near Messina, managed to re-conquer the whole city, took Milazzo, and, after a useless truce, imposed by British and French military fleets only to try to let the rebels to have time to reinforce their army, he restarted the military actions conquering Catania. By May 1849, Sicily was under his control, even if at the cost of much bloodshed. Filangieri was then nominated «Luogotenente generale dei reali domini al di là del Faro», meaning Lieutenant-general of the royal domains beyond the Lighthouse (of Messina).

He knew very well, also because his friend Pietro Calà Ulloa, who had been a judge in Sicily, that a criminal terrible organisation, called Mafia, had revealed its presence in Sicily before and during the Sicilian revolt. He tried to carry out many reforms in the organisation of the State in Sicily. He succeeded in many accomplishments, like building railways, streets, schools and hospitals, and reforming the organisation system of local administrations. However, exploiting their influence on Giovanni

Cassisi, the King's Minister for Sicily, all Sicilian opposers to Filangieri enlightened policy managed to stop his action, also by inventing some fake accusations of malversation.

A second time, disgusted by the dishonest and miserable behaviour of his political opponents, he retired to private life in 1855, not without commenting about his absolute disinterestedness in adding richness to his already enormous fortune. Britain, which was exploiting (without paying the fair economical compensation) the natural resources of Sicily, had managed once more to stop the Neapolitan reforming action in Sicily and to keep exploiting Sicily.

In 1859, the new king Francesco II, following the last advice of his father, who too late had understood the true standing of the man, nominated Filangieri President of the Ministers' Council and Minister of War. However, the king did not allow to Filangieri to nominate the other ministers. Filangieri tried to promoted good relations of Two Sicilies Kingdom with France, to support Piedmont against the Austrians in Lombardy, and to persuade the king about the necessity of an alliance with Piedmont. He also tried to negotiate with Piedmont the federation of Italian States, trying to avoid their forced reunification. Both the proposals for an Italian federative negotiation and for the promulgation of a constitution were refused by the king, so that finally, for the third time, Filangieri gave up and resigned his office.

When, in May 1860, Francesco accepted to promulgate the constitution, Garibaldi and Cavour had already persuaded all Neapolitan and Sicilian centres of powers, by means of bribery or by persuasive propaganda of their nationalistic ideology, to abandon their State. Too late Francesco II tried to call back the now very old Filangieri to try to save his Kingdom: the king arrived to pay himself a visit to the General personal residence, in Sorrento. Filangieri was, however, aware of the fact that nothing could be done, anymore.

Indeed even **Liborio Romano**, the Minister of the Interior appointed by Francesco II, was negotiating secretly with Cavour and Garibaldi, and with Camorra (the Mafia-type criminal organisation which had started to control Naples) to try to get better conditions for the annexation to Italy of the State that, instead, he was assumed to preserve. Liborio Romano, being afraid of the influence which he could have in the subsequent events, before the arrival in Naples of Garibaldi, ordered Filangieri to leave the country. Filangieri moved at first to Marseilles with his wife and then to Florence. He adhered to the new Italian regime, but he refused to accept any official role, peacefully dying in his villa of San Giorgio a Cremano, very close to Naples, on 9 October 1867.

A Reappraisal of the *Res Gestae* (i.e. The Deeds) of Carlo Filangieri

Losers are not considered fairly by history. Winners, to justify their actions, have always the tendency to hide the reasons of those who opposed them.

The history of Italian unification is being re-evaluated only recently, albeit all most important information can be found, for instance, in the very well-documented work by De Cesare: *La fine di un regno* (The end of a kingdom). This documental work was written just after the described events (in the period 1897–1900) and with a wealth of carefully reported reliable sources. It is therefore surprising that only recently historians are reconsidering the events that led to the so-called reunification of Italy. Indeed many are now talking about a «conquest of the South», i.e. a conquest of the Kingdom of the Two Sicilies.

This kingdom was the largest, probably the richest and surely the most attractive region of the Italian Peninsula, as the interest paid by Britain in its exploitation demonstrates clearly. Many important details of the events, which led to its incorporation into Italy, need still to be clarified and understood. We do not believe that some unknown documents need to be discovered, yet, but we could expect many surprises in some archives not completely studied, up to now.

We believe instead that a new spirit must be adopted in describing those crucial events of the European history, which finally produced one of the biggest economies of the planet.

Italy has produced, in his millennial history, great economic and cultural advancements: think about Renaissance, for instance. However, it also gave birth to the criminal organisation, so harmful also in the USA, which we call Mafia.

The narrative imposed by the ideology of the winners, the supporter of Cavour's reunification by violence and bribery policy, describes South Italy as an underdeveloped country under the yoke of a reactionary and obtuse dynasty, i.e. the Bourbons of Naples, which did not care about its subjects. This narrative is contradicted by recent studies about the economy of the Kingdom of the Two Sicilies. Some historical surveys (see Daniele and Malanima 2007) recently published give a scientific basis to the impression which everybody has in visiting South Italy. The impression is that of being travelling through a country which had been very rich and which has, relatively recently, been pillaged. Exemplary is the story of the city of Arpino, which is only one among many similar examples and whose story cannot be considered as anecdotic. Arpino was, during the epoch of the Kingdom of Two Sicilies, a rich city, whose wealth was based on its textile and musical instruments industries. Of course, if one believes that industrialisation is made possible only by the presence of «electric power» plants, he could decide that Arpino was not an industrial city. However, in Arpino, the many rivers, which falls from Apennines, moved the textiles looms. These rivers, in many millennia, had been bridled to supply the needed energy. One could say that city of Arpino was already fighting against the world pollution and temperature raise even if the textiles produced in Arpino were enough to respond in a great part to the demand of the whole Kingdom of Two Sicilies. In Arpino, the natural resources made his inhabitants wealthy and peaceful since the ninth century before Christ, when it was a Greek colony, until Italian reunification. Since 1860 it is a sort of ghost city, where only in summertime those who emigrated (towards USA, Canada, Germany, North Italy and, for the luckiest, the closest Rome) and their descendants can enjoy the pleasant climate and the beautiful landscapes.

Piedmont entrepreneurs took all looms and transported them away, and the city, with his famous Classical Lyceum (High School in Greek and Latin Languages), his Theatre, his Library, his beautiful Churches, Castle, Archeological Site, Monasteries and Musical Instruments factories, was suddenly and nearly completely abandoned.

The arguments about the solidity of the theory of the pillage of Southerner human and economical resources are reinforced by reading the carefully documented chronicle by De Cesare and the volume on Carlo Filangieri by Pietro Calà Ulloa.

Ulloa was the very last Prime Minister of Francesco II, the one who followed the king in his exile. Ulloa after the end of the Kingdom dedicated himself to historical and juridical studies and managed to produce and reorder an impressive amount of historical and literary essays. Remarkable enough, among the erudite production of essays by Ulloa, one has to recall those in which he is probably the first juridical scholar denouncing, already in 1838, the existence and the raise of «that criminal organisation based on strong bonds of association between the members which is defined by themselves: *la mafia*». Ulloa was not a profane, in the subject, as he had been the king's attorney at the Court of Trapani. He carefully put forward the conjecture that this criminal organisation plaid a role in the revolt of Sicily against Naples, and that the British Empire tried to exploit this revolt to get commercial, political and economical advantages. More or less as has been done by USA intelligence during the Second World War.

Ulloa is surely faithful to the Bourbons, but he is attentive to the reality of historical facts and events and is a first-hand witness of the last period of the life of the Kingdom of the Two Sicilies. Ulloa contributes to witness the greatness of Carlo Filangieri, recognising him as one of the greatest protagonists of the nineteenth century Italian and Neapolitan history. Ulloa complements and supports the analysis developed by Teresa Filangieri Fieschi Ravaschieri (1902), which has been accused to be a partisan historian, being the daughter of Carlo. Ulloa adds evidence to the opinion that Carlo Filangieri not only restored the unity of the Kingdom of the Two Sicilies by bringing back, with the force of his armies, to Sicily the legitimate government of Ferdinando II, but also that he was so long-sighted to operate for social reconciliation and lower classes emancipation. He attributes to Filangieri the persuasion that criminality can be confronted only via serious support to the economical independence of lower classes and a decided impulse to cultural and economical development. A modern analysis indeed!

Filangieri, after the fall of Napoleon, participated actively in the reconstitution of the Kingdom social structure, being one of the main architects of a successful example of national pacification. This pacification was obtained by amalgamating, in the institutional offices, the state servants with a French formation and those with a loyalist Bourbon formation. Indeed, Carlo Filangieri had above all a strong sense of the State, probably acquired during his formation in Paris. His loyalty, when he served in an office, was clear, and he put in his actions all his political capacities, always trying to pursue the interest of Neapolitan people. He attempted to get for this people the alliances with France and England, as a way for assuring to the State, the People and the Dynasty, the best outcome of the unavoidable unitary turmoil. He tried to find a federal outcome of the Italian «unavoidable» unification process:

he did not succeed to persuade the king and the dominating Neapolitan classes to follow his advice and the negative consequences on South Italy are still visible.

Historians should learn some lessons about Filangieri's failure to use this experience to perform successfully the European «unavoidable» unification process.

Several Nations tried to unify Europe via military campaigns: the disasters were terrible. The political vision conceived by Filangieri for Italy must find its realisation in the process of creation of a unique European political identity.

Being an honest and frank man, having a social and economical status which allowed him to treat as a peer the king himself, Carlo Filangieri was always courteous, but he never flattened his interlocutors: one should not forget that when he was young his had risked his life in a duel for not ignoring an insult directed to his Nation! Therefore Filangieri was not appreciated by everybody and in all situations: not being fond of it, he several times decided to be exiled from power devoting himself to his family interests. However, when he was in charge, he had always only primary roles. A modern state entrepreneur, *ante litteram*, Carlo Filangieri conceived and realised the construction of a big industrial plant of Pietrarsa, which has been defined with reason "one of the most world-famous Neapolitan political, social and economic jewel of the nineteenth century".

A plaque in the museum of Pietrarsa still states: "*So that the Kingdom of the Two Sicilies no longer have to need the foreign labour to manufacture the machines moved by steam and in order to have restituted to us all our capacity of Italic discovery by means of the education of young Neapolitans, this school of student machinists, Ferdinando II founded, in the eleventh year of his reign, while Carlo Filangieri Prince of Satriano, was governing the learned branches of the army.*"

The Talents Needed to Be a Winner: Power Is the Prize for Machiavellian Behaviour

In any English dictionary, one finds the following definitions.

Machiavellian, adjective: Cunning, scheming and unscrupulous, especially in politics or in advancing one's career. Relating to Niccolò Machiavelli.

One has to accept that in the most Machiavellian country of the world (i.e. UK) the adjective is discredited at the point that he has only negative meanings.

We believe that the political philosophy of Machiavelli was not so immoral as his most talented epigones want to make us believe.

Carlo Filangieri, in his political action, was never Machiavellian, unfortunately. He provided, instead, all his efforts to modernise the army, to found the Arsenals, to build streets, bridges and railways, to assure the efficiency of educational and organisation institutions of the State. As Garibaldi approached, Carlo Filangieri was obliged to leave Naples. Ulloa wrote: «In adversities of politics there is only one thing which is really important: that they were not deserved. And he found in France the respect that is inspired by strong convictions and by a life without weakness or stain».

We somehow disagree:

Carlo Filangieri loss his political battle for having kept a rigidly fair and uncompromising behaviour in a society, which was systematically unfair and totally corrupted.

His great human value was vain, as, tending to the best, he did not manage to assure to his people the good. Indeed after reunification, more than 40 millions of Southern Italian were obliged to emigrate, those who remained in their fatherland are suffering because of organised criminality and Naples has lost a long kept cultural, scientific and economical European primacy. Luckily Piedmontese conquerors allowed us to enjoy a full right to citizenship, so that by emigrating, and hiding our Neapolitan accent, as individuals, we could survive the destruction of our country.

Had Filangieri been a little bit less fair, noble and correct, accepting some Machiavellian compromises, probably he could have been a more effective dictator for his people.

Filangieri was an honest dictator, even when facing criminals or predatory foreigner powers. When the political confrontation became unfair, he always exiled himself, refusing to reduce himself to the level of his political adversaries. His fame, among his contemporaries, was crystalline and such remained also when he was falsely, by persons who were conniving with Mafia, accused of malversation. However, his reputation was not enough to save his people by conquering. History does not remember him outside the region that has been his kingdom. His political action, aiming to a more fair federation of Italian states, was vain.

On the other hand, another politician, much less educated, brave and rich, simply by using a political strategy full of cynicism and opportunism managed to found one of the richest State in the world.

We are talking about Camillo Benso, Count of Cavour.

Camillo Benso, Count of Cavour: The Heterogenesis of Ends

Ahi serva Italia, di dolore ostello,

nave sanza nocchiere in gran tempesta,

non donna di province, ma bordello!

Ah! servile Italy, grief's hostelry!

A ship without a pilot in great tempest!

No Lady thou of Provinces, but brothel!

The Divine Comedy by Dante Alighieri, translated by Henry Wadsworth Longfellow, Vol. II (Purgatorio) Canto VI-78

It is difficult, in modern historiography, to find even the slightest criticism about the personality, the actions and the choices of the Count of Cavour.

His figure was sanctified in a secular rite that, in the Italy of the Notables (i.e. the Italy which was shaped by the bloody reunification process governed by Cavour himself), led to the naming after him a street and a square in every big or small town in Italy.

To find these criticisms, you have to dig into the biographies of his political opponents elected in the Parliament of Turin or later in the Italian Parliament. On the other hand, the data obtained from the official biographies shine for the polite omissions: What was the formation of Cavour? Which schools did he attend? Who influenced his culture and his personality? It is very difficult to deduce this information from encyclopedic entries or from the most popular learned biographies. We have dug in the available sources and we have found that a different vision of his personality is possible.

It is useless to report here the list of all achievements that have been attributed to Cavour: most likely it is true that he was the author of the actions that caused the events for which he is glorified. What we claim is that he had no strategic view, he was only an extremely gifted tactician. He did not know what he was doing, but he did it!

He left to Italy big public debts, and he caused a violent civil war in South Italy, the emigration of at least 20 millions of South Italians immediately after reunification and the economical and social collapse of South Italy. But, he managed to get Unity and to give birth to one of the most powerful economical Nations in the world.

To understand better, the social phenomena involving Cavour it is necessary to use the theoretical efforts produced by a Neapolitan philosopher: Giambattista Vico. Indeed, coming to think of Vico's philosophy, Cavour is one of the greatest personifications of one of the most important historical phenomenon described by Vico: we are talking about the heterogenesis of ends (often translated also as heterogony/heterogeneity of goals/purposes). As clearly explained by Vico, very often one can observe «conseguenze non intenzionali di azioni intenzionali» i.e. «non intentional consequences of intentional actions».

The Heterogenesis of Ends is the social phenomenon, which can be observed when an ongoing sequence of actions and decisions occurs under ever-shifting patterns of primary and secondary goals and finally produces a completely unexpected output.

Italian reunification is possibly one of the most grandiose of such casual sequences of actions and decisions.

Consider, for instance, that it is historically proven that Cavour wanted to stop Garibaldi in Sicily, incorporate Sicily in the Sardinia Kingdom and then find an agreement with the Bourbon King. To this aim he even sent some weapons to Liborio Romano, the Minister in charge in the government of the Kingdom of Two Sicilies, to help Naples to resist to the army of Garibaldi. Instead, Romano betrayed both his King Francesco II and Cavour, as he desired to become an important politician in Unified Italy and replace Cavour. To this aim he helped Garibaldi to conquer easily Naples.

Coming back to the Unity of Italy, the question is: such an achievement could have been done better?

Two related questions are:

Why the dominant classes of the Kingdom of Two Sicilies allowed Cavour to conquer their country? Why Carlo Filangieri did not manage to get better conditions for the unification of his country to the rest of Italy?

The answer can be given in a brutal way. Filangieri had an ethically high ideal, was pursuing it with a long-sighted vision and was not distracted by tactical diversions necessary to get some immediate economical advantage or personal gratification. Cavour, instead, was not inspired by great visions: in every circumstance, he looked for the immediate advantage for himself or for his group of power. Filangieri wanted to realise great and visionary revolutions, having been dazzled by Napoleon. Cavour seemed instead, in every moment of his life, to pursue the possible small (sometimes miserable) advantage here and now. This behaviour eventually prevailed.

Cavour was never recognised as a master of eloquence.

Angelo Brofferio, a member of the extreme left party in the Piedmontese Parliament, openly opposed Cavour's policy. In particular, Brofferio opposed Cavour's monarchical ideas and his bills aimed at achieving Italian independence through the Piedmontese involvement in the Crimean War and in the political-economic alliance with England. Recall that while Filangieri invited the Russian autocrat to visit the industrial plant of Pietrarsa, in order to help him to establish a similar one in Russia, Cavour sent to Crimea a relatively big army, to fight against Russia, and to support the imperialistic actions of England and France.

   Brofferio said about Cavour's political conduct:«he does not have a precise political orientation, nor respect for conventions and morality...».
   Moreover while speaking about Cavour's personality, culture and reputation, he added (the very elegant Italian by Brofferio is very difficult to translate: we will try our best to keep his elegant form and an intact meaning):

> *It was harmful for him his plump figure, his vulgar appearance, his ignoble gesture, his ungrateful voice. Of literacy, he had no trace; of arts, he was profane; of every philosophy, he was ignorant; any ray of poetry did not flare in his soul; he had a very little education; the words that came out of his mouth were all mumbled with his French-like accent* [Probably Cavour was never capable to speak correctly Italian, as he was using a kind of French based dialect, common in Piedmont of his epoch: here Brofferio does not want to say, as some apologetic historians tried to suggest, that Cavour could speak only French. Cavour could only speak fluently the dialect of his village. In this aspect Cavour resembles Don Pasquale, the main character of another part of this book]; *of his solecisms there were so many that to agree them with the dictionary of the Italian language, it would have seemed an impossible task for everyone.*

Historiography, even the most hagiographic, agrees with Brofferio when stating that Cavour did not have a unitary and precise strategy. Cavour did not know, probably, what strategy actually is. He was, however, a great *Maestro* of tactics. With certain variations, almost all Cavour's biographies and essays on his political action contain statements of this kind:

«Although he had not formulated a pre-ordained project of national unity, nevertheless Cavour succeeded in managing the historical, economic and political events that led to the formation of the Kingdom of Italy.»

A Short List of the Achievements of Cavour

His complete name, including his nobility ranks is Camillo Paolo Filippo Giulio Benso, Count of Cavour, Isolabella and Leri (10 August 1810–6 June 1861). However, everywhere in Italy, he is simply called Cavour.

Il Trasformismo.

Cavour was one of the leaders of the Historical Right, the conservative moderate party in the Piedmont Parliament. However he conceived, with Urbano Rattazzi (the leader of an apparently more progressive Italian party), the policy of «trasformismo».

This concept cannot be explained in English easily. There is no word with a similar meaning developed in any English-speaking Parliament: in AngloSaxon culture this kind of policies are practiced, but never openly declared. Instead in Italy, the capacity of manoeuvring and manipulating is a reason for being proud and many Italians happily confess their capacity of deceiving.

The word «trasformismo» can be rendered into English only via a circumlocution. For instance, one can use the following one: «the action of shifting alliances to suit political needs». Rattazzi and Cavour found a series of political agreements, tactically assuring, to each of the two groups, which governed together, the partial satisfaction of several electioneering demands. As a consequence, they produced some bills and laws non-completely coherent one with the others and caused some serious budget problems to the Piedmontese State. Those who believe in the recursive structure of social events, as postulated by Vico, will find astonishing similarities between political situation in Italy during 2019 and in Piedmont during 1852: we leave to the reader to imagine who is playing the role of Rattazzi (the leader who lost his power in this political game) in 2019 situation.

Cavour supported his political ascent by founding the political newspaper *Il Risorgimento,* and he was elected deputy, at the elections of 1848, to the Chamber of Deputies. His Machiavellian conduct allowed him to rise quickly in rank through the Piedmontese political establishment and, via the described practise of «trasformismo», he arrived to easily dominate the Chamber of Deputies. After having supported a very expensive railways system expansion program, Cavour was elected Prime Minister of the Kingdom of Piedmont-Sardinia.

This position he managed to maintain (except for a six-month resignation) during a very turbulent historical period: the Second Italian War of Independence and Garibaldi's campaigns to unite Italy. He was nominated on 4 November 1852 and resigned on July 19th 1859, because the King had signed the peace treaty with Austria without daring to consult him. It seems that Cavour, in this occasion, shouted, insulted the king and, after having destroyed a chair, went away literally slamming the door. However, the king was obliged to re-nominate him on 21 January 1860, as no Prime Minister could get the confidence of the Chamber. He maintained the office until when the Kingdom of Sardinia became the Kingdom of Italy. He kept the position of Italian Prime Minister between 23rd March 1861 until his death on 6th June 1861. Therefore Cavour did not live enough to see Venetia or Rome incorporated into the new Italian Nation.

Without having available the necessary economical resources, Cavour pushed the Kingdom of Sardinia to adopt several reforms and development programs, thus leading his State nearly to bankrupt: only with the treasure pillaged from the Kingdom of the Two Sicilies, the public debt of the Kingdom of Sardinia could be diluted.

This is not a biased opinion of some Bourbon nostalgic supporter: during the session of 4 March 1861 of the British Parliament, the Irish Conservative and Catholic deputy John Pope Hennessy (1834–1891), accused openly the British government of having favoured and financed the revolutionary party in South Italy. In fact, in the Parliament minutes one can read, «the Piedmontese conquest of the Peninsula was inspired by the little noble reasons to resolve the serious financial crisis that gripped the kingdom of Vittorio Emanuele [of Savoy] with the acquisition of the resources of the other Italian States, which were all in a more prosperous economic situation».

The Machiavellian attitude towards political action was fully deployed by Cavour, when as Prime Minister, negotiated the participation of Piedmont in the Crimean War, thus obtaining the support of France and Britain for a war against Austrian Empire. In this way the Second Italian War of Independence could be partially successful: indeed it was interrupted half-way by a second thought of Napoleon III. In the next period Cavour, more or less, controlled Garibaldi's expeditions, trying continuously to stop him, being afraid of his true intentions: Cavour always suspected that Garibaldi wanted to establish a Republic instead of a Kingdom, in the unified Italy. At the end, thanks to a series of unintended manoeuvres, he successfully managed to transform Piedmont into a new great European power, obtaining the annexation of territories which were five times larger.

Dictator or tyrant? The big Italian public debt.

A remark is here appropriate about the tendency of the dictator Cavour to become a tyrant: the English historian Denis Mack Smith admits that Cavour has been, probably, the most successful leader in an Italian Parliament, ever. In this he forgets, probably, De Gasperi and ignored that Cavour died rather young: he did not have the time to ruin his masterpiece. One can, however, imagine what he could have done later if he thinks about the action of Giolitti, who was clearly a follower of Cavour, especially in his disastrous policy of «trasformismo». Giolitti ambiguous leadership opened the way towards power to Mussolini: Giolitti, having the majority of the votes in the Parliament allowed to this true tyrant to reach the power. Cavour, always according to Mack Smith, was too often dictatorial, his action simply avoided taking into account the opinion of his ministers and too often he overpowered the privileges of the Parliament. The most dangerous anti-democratic action enacted by Cavour, who showed "how to do" to too many of his successors, was his interference (i.e. his vote-riggings) in parliamentary elections: as Prime Minister in charge he manipulated the election of deputies in order to control easily the majority of the Chamber also after a new vote.

Cavour is the precursor of another big problem, which has afflicted Italy during the whole of his modern history: he can be regarded to be the "father" of the enormous Italian public debt. The Kingdom of the Sardinia, before reunification, not only had a public budget in total disarray but also, because of the habits established by Cavour, systematically hid the true amount of the public deficit. Instead, in South Italy, the Kingdom of the Two Sicilies was exerting a remarkable budgetary rigour. Among the many scholars who witnessed this situation we must mention the Neapolitan economist Giacomo Savarese who was a direct spectator of the events that led to unequal reunification of the public debts of the two bigger pre-unification Italian kingdoms.

We do not want to say that the more than 2300 billion euros of debts of the Italian Republic (in 2018, later who knows what will happen!) are directly inherited by the debts of Piedmont. We simply want to remark that the bad habits of «trasformismo» (i.e. the habits of making everybody happy by spending public money and postponing indefinitely the payment of the debts, which are transferred to the future generations) are not habits, which were transmitted to us by the «badly» organised Bourbon State. It was the dishonest and unscrupulous management of public accounts started by Piedmont which has, for more than a century and a half of Italian history, maintained that mark, thus establishing a very uncomfortable cultural inheritance.

The Neapolitan Economist Savarese documented in his most important essay (Savarese 1862) that Piedmont had its finances completely out of control: indeed its public debt increased more than 565% in the short period including the ten years before the unification. It was Cavour that started the policy of cheerful finance. It was the Parliament of Turin, under the control of Cavour, which established the Italian standards and culture of easy debt.

The Kingdom of Sardinia had only one way to easily get rid of the burden of its debt.

The so-called «wars of independence» and the (re-)unification of Italy were probably the simplest (but also the bloodiest) way to settle Piedmont public budget, whose large deficit was due to an adventurous investments policy and to the reckless expenses for nurturing the «trasformismo» system of clienteles.

Also, the fable of a productive North Italy must be put in its right perspective: indeed Piedmont had a trade balance constantly in red (see again Savarese, but also Nitti 1903) and it is indubitable that these enormous debts were not only due to the reckless financing of its wars. In reality, wars were financed to try to resolve the problem of an already enormous debt, with the immediate consequence, however, of increasing it. The start of this «financing policy» was the Crimean expedition, which is the intention of Cavour was needed to focus the attention of the France and England on the "Italian question" and which, instead, was useful only to make economically convenient for these powers the war against Russia: transferring on Piedmont, and finally on South Italy, the charges to produce the British and French profits. Remark that, wisely, in the same period Filangieri had started a peaceful collaboration with Russia.

To be more specific: the Piedmontese debt amounted to 168 million lire, in 1848 when Cavour started his dictatorship, to arrive, in 1859, to 1.12 billion lire (exactly the said increase of 565%), which was equal to almost 74% of the GDP. Piedmont was paying as annual interest on the debt the amount of 68 million lire, more than three times the amount paid by the Kingdom of the Two Sicilies.

In the South, the economic productivity was much greater and more diversified, as the anecdote about the city of Arpino strongly suggests. The GDP of the Bourbon kingdom, the years after reunification, amounted about to 2.62 billion lire, (a *pro capite* yearly income of about 291 lire) while that of the Piedmontese kingdom to 1.61 billion lire (a *pro capite* yearly income of about 220 lire). In nearly the same period (1847–1859), the debt of the Two Sicilies had increased by 29.6% to 411.5 million lire from the initial 317 million lire. We have not seen any serious economist to doubt about these numbers: consider also that Nitti and Savarese, who are famous and reputed economists (Nitti was also Italian Prime Minister!) are not the only historians and scholars reporting the listed figures. A huge amount of sources and data confirming their analysis if given also by De Cesare.

It has to be remarked that also De Cesare was deputy to the Italian Parliament: the fact that civil rights were assured to the individual citizens of the conquered State had this drawback: some of them continued to complain for a long time because of the injuries inflicted to their Country, even if the majority of those who emigrated towards north tried, in general, to hide their origins.

Therefore, the debt to GDP ratio of the Kingdom of the Two Sicilies was 16.57%. The reader is invited to compare this value with the corresponding rate in 2018 for Germany!

Because of the wealth generated by its healthy economy, from 1815 to 1848 the Bourbon kingdom could flourish imposing only five taxes which produced an increasing amount of public revenues (from 16 to 30 million ducats) because of the increase of the GDP. In the time interval between 1847 and 1859, no new tax needed to be introduced and the public budget was decidedly so transparent that the Neapolitan debt was considered, in Amsterdam, one of the most reliable investments. In the Two Sicilies, no sale of state property had been ever made.

Instead, the Kingdom of Sardinia obscured all the data on public debt and even stopped calculating the state budget while inventing many new taxes and starting selling state property. Francesco Nitti documented that Piedmont had assets of 27 million lire of gold while for the Kingdom of the Two Sicilies these assets amounted to 443 million lire of gold: Piedmont was strangled by Rothschilds, and it is suspected that this family supported the conquest of Naples to get back their money. Indeed the only thing that Cavour could do, to avoid bankruptcy, was to conquer at least Sicily, Lombardy and Tuscany. Naples was added to the booty by the foolhardiness of Garibaldi.

This analysis led, in 1862, Savarese to talk about a "tragic inheritance". It was confirmed by Vittorio Sacchi, a (Piedmontese!) economist who was charged by Cavour to asses the state of the public finances of the Kingdom just annexed.

Vittorio Sacchi was famous for his honesty and was imposed to Cavour by the Senate. He was a competent magistrate who had controlled the finances of many institutions of the Kingdom of Sardinia. He also had been a brave soldier who got a medal for his participation in the 1848 war against Austria. He was for three years the director of the Banco di Napoli, one of the two National Banks of the Two Sicilies Kingdom. Remark that the other National Bank, the Banco di Sicilia, had been pillaged during the occupation of Palermo.

In his report, Vittorio Sacchi wrote

Nei diversi rami dell'amministrazione delle finanze napoletane si trovano tali capacità di cui si sarebbe onorato ogni più illuminato governo

In the different branches of the administration of the Neapolitan finances one finds so high capacities that most enlightened government would be honoured by them

and for this reasons he suggested to his Prime Minister (Cavour) to adopt without procrastination the administrative mechanisms of the Kingdom of Two Sicilies.

Obviously, Cavour ignored Sacchi's suggestion.

Phenomenology of Dictatorship

The word «leadership» is universally associated with a positive meaning. In the previous pages, notwithstanding the unavoidable criticism, which was necessary to comment their actions, we have described heroes whose actions had, eventually, an overall positive effect on the group which had entrusted them.

Now on, instead, we will discuss the behaviour of personalities whose actions, we believe, had an overall negative effect on their people. Therefore, we decided to use, with a negative meaning connotation, the word «dictator».

The Weaknesses of Two Sicilies as Exploited by Cavour and Garibaldi

Of course, if a wealthy country, with a very well-organised economical system and a strong economy, is pillaged by an economically weaker country, then it must have some weaknesses in its cultural and social structure and traditions.

This is another point that should be carefully studied by a General Theory of Societies and Nations. Here, we simply recall that to be a rich, economically strong and scientifically advanced Nation is not enough to be protected from becoming prey of less advanced ones.

An ancient example of poorer State conquering a richer State.

We open here a parenthesis by recalling here the events, which led the Roman Republic, by means of his consul Marcello, to subjugate Syracuse, a much more economically and technologically advanced, rich and militarily strong power.

The Romans after having lost all the battles managed to win the war. Why? How? In one word: Treason!

We will, however, spend more words, as what happened in the confrontation Rome versus Syracuse can be paralleled with the later events leading to the reunification of Italy.

© Springer Nature Singapore Pte Ltd. 2019
F. dell'Isola, *Big-(Wo)men, Tyrants, Chiefs, Dictators, Emperors and Presidents*,
https://doi.org/10.1007/978-981-13-9479-9_6

Rome had developed a relatively more efficient political system, with a consuetudinary constitution (this is a situation which can be very effective, as it is also shared by the British Empire!) that assured, for a long time, the efficiency of the Roman political institutional system. The Roman sophisticated legal structure included detailed procedures for approving, enacting and enforcing the laws and an efficient judicial system. In Rome, what has been called «the certitude of the rule of law» was considered a «sacred» social value. The reader will contextualise this statement: we are aware of the fact that too often the political fight in Rome ended will the homicide of the political opponents or with Civil Wars. However, the Roman legal system somehow limited the use and abuse of violence, in the process of the establishment of the social choice. In the spirit of Condorcet and Montesquieu, the Roman system was an important step forward towards a modern legal system.

Syracuse, instead, like nearly all Hellenistic States, was simply a tyranny. The populist democratic system which was used in Athens was not structured enough to persuade the Hellenistic States to adopt it as an efficient way for organising wealthy societies. Hiero II, the tyrant of Syracuse, was probably supported by Archimedes, but had many political internal enemies. Some Syracusan citizens preferred to sell their State to a foreign enemy rather than supporting Hiero. The institutional system of Syracuse was intrinsically weak, and notwithstanding its overall social and cultural superiority, it became subject of the less scientifically and technologically advanced Roman State. The reason is simple: Rome, with its institutional system, did manage to get a greater social cohesion and identity. *Civis Romanus Sum*: I am Roman Citizen, has been a proud statement, which had a tremendous importance for centuries. Every Roman citizen identified himself with the power of his State, and, even if pursuing his own personal interest, he always had a great respect for «Roman legality».

Piedmont was a constitutional monarchy, and Kingdom of the Two Sicilies was an absolute monarchy.

Coming back to the Italian reunification process: Piedmont had endowed itself of a written constitution (Lo Statuto Albertino) and a set of institutions (Deputy Chamber, Senate, a government having the confidence of the Chamber, etc.) assuring to his social structure a stability and an efficiency unparalleled in pre-unitary Italy. Instead, the bigger, richer and economically well-organised Kingdom of the Two Sicilies never managed to attain a solid social organisation and reliable institutional structure. Even if his philosophers envisioned very early the need of establishing a constitution, even if the University of Naples, founded by the visionary Frederick II Hohenstaufen, did express the strongest school of law in Europe, the Two Sicilies had a frail and fragile social structure. Everybody with a public office in that State pursued, without scruples, his own interest and that of his family, without worrying (at all!) about the interest of the Nation.

The Neapolitan people were eventually abandoned in the hands of the new Italian regime and, in an extreme effort to rebel to its destiny, organised a guerrilla which has been disparagingly called «Brigantaggio», i.e. brigandage.

Those Masters of Duplicity Who Surrendered Naples

We will describe only briefly, in what follows, the despicable behaviour of the Neapolitan Generals Ferdinando Lanza and Francesco Landi. Instead, we will give some more details about the deeds of the infamous Neapolitan Minister Liborio Romano.

However, the list of traitors of their country could be very long. Nearly, all of them did not get any of the expected advantage because of their treason: only some happy few were really rewarded, those who managed to reach power position in the new Kingdom of Italy or managed to keep their ancient privileges, as the Leopard.

Don Liborio Romano has been an unsuccessful Neapolitan double traitor. Actually, double is not enough to qualify him, as we will see that a most appropriate epithet should be «multiple traitor». Obviously, his contribution to the heterogenesis of ends leading to the unification of Italy has been erased by the Piedmontese winners and, in particular, by Cavour. Don Liborio is like a human character that could be paralleled to the Pulcinella in the Italian «Commedia dell'Arte». He personifies the man who sold his king, the people whose destiny was entrusted to him, and his country to a ravenous enemy, without managing to get nothing more than a very small and ephemeral miserable personal advantage.

On 7 September 1860, he erroneously believed to be a true genius of political manoeuvres, managing to parallel those performed by Cavour. He was sitting together with Garibaldi in the carriage that led the general to his triumphant entrance in Naples. Don Liborio wrote in his memoirs: «And Garibaldi, a sublime and indescribable spectacle, entered Naples, [..] calm as if he were returning home, modest as if he had done nothing to get there!».

Don Liborio knew, instead, that Garibaldi had paid (or promised to pay) a lot to be able to arrive there! Don Liborio probably never understood that he had not lived a moment of well-deserved glory but had sealed his historical damnation, as he had been nailed, because of his underhand actions, to the role of the stupid turncoat.

This is the true difference between the turncoats from North (as Cavour) and the turncoats from South (as Don Liborio).

The first ones manage to increase their reputation because of their manoeuvres, while the second ones exaggerate and finally manage to ruin it.

Don Liborio was born in a small Apulian town, and his life has been a continuous and clamorous self-contradicting sequence of decisions and actions.

He personified the «trasformismo»: but in the Neapolitan way.

He started as a young member of the anti-Bourbon party, but then he manoeuvred to be appointed by Francesco II as Minister of the Interior, making the king believe that, holding this office, he was bound to behave in a loyal way. The appointment was made on 14 July 1860, when Garibaldi, who had landed in Marsala on 11 May, already occupied a considerable part of the Kingdom of the Two Sicilies. His family boasted to descend from a branch of the Romanovs, and he became famous for his reputation as unscrupulous lawyer: he had the audacity to defend the interests of the British Empire against those of the Neapolitan State in a process. He managed to inflict to his own country costly economical compromises. To cover his greed for personal interests, he pretended to have liberal ideals. In this way, he could justify his imprisonment and exile, which probably has been inflicted to him for less noble reasons. When he was finally allowed to return home, he was kept under strict surveillance by police, even if he was allowed to continue his forensic activity and treated with great concern because of his solid relationships with organised Neapolitan criminality: Camorra.

Romano, as a Bourbon Minister, showed some capacities in producing effective propaganda: for instance, he had posters put up nearly daily to inform the people about his initiatives and (*who knows?*) even possibly to send coded information to his criminal allies. Indeed, as Minister of Interiors, he showed the true extent of his unscrupulous political duplicity. This last has to be compared with that showed by Cavour. Undoubtedly, Cavour was much more effective in betraying: he practised treason only when a decisive advantage was obtained, otherwise behaving in an apparent fair way. Of course, the Napoleonic correctness of Carlo Filangieri was bound to succumb against the combined action of these two sharks.

Notwithstanding and because of his office, Don Liborio played a political game on his own. He was not simply two-faced: he managed to be even four-faced. First of all, he was Minister of Francesco II; therefore, he had explicitly accepted to be loyal to the *status quo* of the Kingdom of the Two Sicilies. However, as soon as he had taken into office, he immediately started a secret correspondence with Cavour and, simultaneously, established good relations with Garibaldi. In this way, he believed to be «covered» both in case of the prevailing of the Garibaldi republican cause but also if the Savoy would have managed to dominate the whole Italy. Last but not least, Don Liborio supported Camorra for many equally cogent reasons: to be ready to use it in case of a dethronement of Francesco II without loss of the independence of the Kingdom of the Two Sicilies and, in any case, to have a sure «popular» support in case of elections in the new (may it be Italian or Neapolitan) parliamentary regime. To play his spy game, Don Liborio managed to have a telegraphic device installed in his cabinet, circumstance that raised many suspects in the Neapolitan government and court.

The efficacy of his «popular» network has been fully tested: in the first Italian elections, Liborio Romano succeeded to be elected in as many as eight colleges! No other candidate, in the whole Italy, was elected in more than three colleges: Cavour included. In South Italy, history has proven that this kind of performances is impossible without the support of organised criminality.

We shortly interrupt here, for a while, the description of the ominous feats of Don Liborio for sampling those of two other important generals of the army of Francesco II. The reader will find the detailed sources about these two Neapolitan figures in the monumental work by De Cesare. In this masterpiece of historical chronicles, very careful and detailed, one will find that we are not talking about anecdotal behaviours, but that nearly every member of the Neapolitan dominant class (with the exception of Carlo Filangieri and few others) shared them.

General Ferdinando Lanza.

When Francesco II understood that Garibaldi was landing in Sicily, he asked, too late (once more!), an advice to Carlo Filangieri. He indicated the General Ferdinando Lanza, who had been his collaborator in many campaigns. As many other human beings, Lanza was a chameleon: honest and brave, if surrounded by honest and brave people, while dishonest, coward and corrupted when in contact with this last kind of persons.

In the following chapter about Nash equilibria, we will talk about «mixed strategies» used for «winning» games: Lanza personifies the kind of players that behave honestly, at first, to get a personal advantage becoming dishonest, when he was given the possibility.

Francesco II nominated as his Viceroy in Palermo the General Lanza. De Cesare relates the first action decided by Lanza once arrived in Palermo: to send the Vice-Royal carriage (with all the horses) back to Naples. He had, indeed, immediately discovered at his arrival that his privileges included the possibility of disposing of this carriage. He lost some important hours to send a telegraphic message to his eldest son, for giving him the needed instructions: to sell their own carriage and horses and to go the harbour of Naples to receive the Vice-Royal ones: a miserable behaviour indeed.

Moreover, Lanza chose the General Landi to try to repel Garibaldi in Marsala and «was wrong» in giving the right instructions to the army which was arriving from Gaeta to reinforce Landi's troops. Instead of sending them close to Marsala, he ordered them to arrive in Palermo, so helping Garibaldi to reinforce his stronghold in Sicily. Finally, without any apparent reason, he ordered to cannonade the city of Palermo so killing 600 innocent citizens and causing violent riots in favour of Garibaldi. De Cesare could not find any evidence of bribery for Lanza: however, the *vox populi* in Naples claimed with certitude that Lanza had been corrupted by Cavour. The popular belief was reinforced when Lanza, during his Neapolitan dictatorship, went to homage Garibaldi respectfully, in his residence of Palazzo d'Angri.

Finally, when the troops of the Two Sicilies were still at least three times more numerous than those of Garibaldi, Lanza decided to abandon Sicily in the hands of the revolutionaries. The really important deeds had been accomplished: the Vice-Royal carriage was already in Naples! De Cesare reports that while Lanza was reviewing his troops before departing from Palermo a soldier shouted «Eccele', verit' quant' simm. E amma' fui' accussì?» (the Neapolitan statement can be translated as follows: «You Excellency, look well how many we are! Why should we escape in this way?»). Always following De Cesare, Lanza answered: «Be quiet! You are drunk!». Some gossips in Naples reported that Lanza had received a big amount of money in a foreign bank. It is, however, sure that he abandoned the huge treasure of the Bank of Sicily in Palermo. Useless to say, this treasure disappeared quickly and after reunification it was very difficult to discover who had taken it: some Italian Parliament Committees tried to investigate to unveil the mystery, without managing to succeed.

General Landi succeeded in a unique glorious venture.

Having troops amounting to three thousand elite soldiers perfectly equipped and loyal to the cause of the King Francesco II, he managed to lose a battle, in Calatafimi, against about one thousand irregular soldiers, outgunned and poorly organised. Calatafimi was a decisive event, announcing for the end of the Kingdom of the Two Sicilies. On his return to Naples, Landi and many other officers were judged by an investigation commission, which acquitted all the accused: nevertheless, even if he had been judged innocent, he preferred to be discharged. After few months, in the city of Naples the rumours were spread that the former general had tried to cash in the Neapolitan branch of Banco di Napoli a bill of exchange amounting to 14,000 ducats of gold, as a reward received by Garibaldi. However, the bill of exchange had been falsified, since, in reality, its true value was only 14 ducats, three zeros having been added by an unknown hand. The story caused an enormous scandal: possibly already while discussing with the cashier of the Banco di Napoli, Landi had a stroke, which finally led him to death. The scandal was so big that one of the general's sons tried to save the family's honour, by demanding and obtaining a letter from Garibaldi himself, which denied the whole story. However, four of Landi's five sons, which all had been former officers of the Bourbon army, were suddenly nominated officers of the Savoy army. Only the fifth son, Francesco Saverio, remained loyal to his king and died very young in the battles of the Volturno in October 1860.

Let us come back to Don Liborio.

While Garibaldi was advancing towards Naples, the Machiavellian Cavour tried to use Don Liborio to stop him. In the correspondence sent to Don Liborio, Cavour thanks him for "his enlightened and strong patriotism" and for his "devotion to the cause" of Italian reunification. Double crossing in this way Don Liborio risked his head: when Francesco II abandoned Naples, as he refrained to oppose Garibaldi with a battle which risked to destroy his beloved capital, he warned Don Liborio by saying, in Neapolitan language: «Don Libò uardateve 'o cuolle», which can be translated as «Don Liborio, be careful with your neck», i.e. *be careful, because with your behaviour you risk to be decapitated*. Remark that Francesco II used, in a bitterly ironical way, the appellative «Don» for Liborio Romano, as this word, uttered by a king in this way, was extremely unusual. Remark also that Camorra bosses are often called with the appellative «Don». The following chapter will exemplify its meaning, in South Italy.

Indeed, the reader will compare, in the following pages, the behaviour of Don Liborio with the similar behaviour of Don Pasquale, the prototype of a similar kind of dictator. We agree with the mother of Don Pasquale: probably, he would have been a better politician than Don Liborio. Moreover, those who know Don Pasquale will find a striking physical resemblance with Don Liborio, after having seen the pictures of both of them. The common feature is the conniving, duplicitous and deceitful lights, which sparkle from their eyes and their expressions. Also, they both try to justify their despicable behaviour pretending to have tried to pursue noble ideals. Both connive with different expressions of Camorra, playing the role of «well-educated» members of this organisation.

Don Liborio Romano had secretly promised Cavour to persuade Francesco II to leave Naples.

To fulfil this promise was, indeed, very easy as Francesco II had already decided to try to resist in Gaeta, which had always been a military stronghold, as he intended to limit the lives' lost among his people. By the way, Gaeta, which before the reunification has been a rich and culturally active city, the past capital of a Longobard Duchy, since 1860 is reduced to a sort of «big» village, as a consequence of the violent battle occurred there, between the Piedmontese and the Neapolitan armies.

With Cavour, Don Liborio agreed to form, immediately after the departure of the king, a provisional government, and this government had to formally request Piedmont to land his troops from some ships that had been sent purposely in the bay of Naples. The intention of Cavour was clear: he wanted «to save» Naples from Garibaldi.

The second twist by Don Liborio.

Showing how unreliable he was, Don Liborio preferred to ask the help of Camorra to control Naples, while Garibaldi raced North with a small amount of soldiers: for a while, Don Liborio played the role of supporter of the radical revolutionary party! How did he manage to arm the «Camorra» troops? Using a load of rifles sent by Cavour to Don Liborio for helping to stop Garibaldi. This load was disembarked by many «camorrists».

Even in Italy, the country of explicit and boasted «trasformismo», this cynical continuous series of twists was too much.

Later, in the Italian Parliament, nobody wanted even only talk with Don Liborio. This wise ban, unfortunately, was not applied against the Italian continuators of his policy of compromise with criminality, as history of unified Italy sadly proves. To try to recover his reputation, as a member of the Italian Parliament, Don Liborio wrote a long memorial in which he described the critical situation of South Italy, as caused by the improvident annexation of which he had been accomplice.

Don Liborio, after having helped Camorra to pillage the city of Napoli, during his short office as Bourbon Minister, while being a member of the Italian Parliament condemned the:«illicit enrichment that had taken place after the unity when incorporating the Bourbon financial instruments». Historiography agrees in stating that Don Liborio was not capable to act in a democratic Parliament as he lacked any attitude to mediations. He wanted, with myopic vainglory, to be elected in eight colleges, instead of supporting seven other candidates loyal to his policy, as he did not understand the importance of being part of a political parliamentary group.

Don Pasquale

(Don is the ancient Italian word used as a short cut for the Latin dominus, i.e. Seigneur)

Les habiles tyrans ne sont jamais punis.
Clever tyrants are never punished.
Voltaire Mérope, Act V, scene V (1743)

An advanced society must learn to recognise and use the talents of its citizens. One of the main tools of the future general theory of societies will have to allow a characterisation of the dictatorial mind, to direct it towards its best social use. The waste of Don Pasquale's gifts cannot be tolerated in the societies of the future.

A Gifted Dictator.

The young Pasquale was a twin (not homozygous) to an infant prodigy. From the very first years of his life, his brother Antonio explicitly showed an incredible ability to understand numbers. Pasquale, instead, about numbers, equations and logics did not really understand anything: never in his life! The involuntary competition with his brother made him suffer a lot. His father did nothing to hide a clear preference for Antonio, whom he praised and supported continuously. Pasquale considered this preference very annoying. He began to hate his father and his brother deeply. This hatred became only a little less angry when Pasquale realised that he had a superior dowry, too.

Pasquale, in all his life, could always understand the human soul.

To be precise: Pasquale reads people's minds as if nothing could be hidden from him. A brief conversation was enough for him to understand the intentions, weaknesses, aspirations and ambitions of his interlocutor. He had inherited this dowry from his mother. She tried, in vain, to explain to him that, in the eyes of the world, the father had no preference: only the two of them could see this preference, and only because they could use their shared superior gift. Despite the help of his mother, Pasquale never managed to forgive his father. He went so far as not to speak for thirty years with his eldest son who was named after his grandfather and who had inherited from his grandfather the enormous respect for mathematical intelligence together with the total inability to read in the human soul. Pasquale decided to humiliate his father and brother. He would become a great mathematician, a famous mathematician. He carefully investigated the subject, and at the end, he managed to understand what a mathematician is «in practice».

A mathematician is, for Pasquale:

1. Author of many publications in international journals with a peer-reviewing editorial board.
2. Cited many times in MathSciNet.
3. A professor holding a chair of mathematics in a prestigious university.
4. Member of many editorial boards and possibly editor of a prestigious scientific journal.
5. Supervisor (maestro) of many mathematicians who can verify the criteria from 1. to 4.

6. Member of, at least, a prestigious scientific academy.

He often heard, in his life, that a mathematician should have original ideas and great intellectual creativity: but he never managed to understand what these things really meant.

He worked, effectively, all his life to verify the conditions just listed: and, at the end, he became a «famous» mathematician. His mother tried to persuade him to invest his time to become a politician: she rightly believed that this job was more suitable for his son. However, for Pasquale it was only what his father considered important to be really important. His mother was forced to see her son burning his chances of becoming Prime Minister in a neurotic attempt to indulge and, at the same time, humiliate his father. As Pasquale was not very venal, he was able to reach compromises and he knew how to understand the true intentions of every human being; we agree with his mother. Pasquale would have been a perfect Christian democrat: a worthy heir to Andreotti. His mother immediately understood that Pasquale could foresee all the future actions of any person he could meet, in any contingent situation. And she knew very well that these all are the qualities of a true leader and not of a true mathematician.

We said: all the actions of any person? Actually we were wrong!

Pasquale met his future wife by chance. And he immediately discovered that she was the woman of his life. And for a specific reason: neither when he met her and never in their common life, he was able to understand what the true intentions of that woman were. Never!

The Difficult Beginnings

Pasquale was forced to refine his control technique since his primary school. First of all, he made his brother believe that he was really fond of him. The brother diligently did all Pasquale's homework for him until his high school diploma and prepared him for every possible question, which could arise in oral tests. Instead of being grateful, Pasquale hated him more and more as his debt increased. But the university admission examinations at the time of Pasquale were difficult, very difficult. And Pasquale was forced to ask to be received by the president of the examining commission. He managed to do so thanks to an acquaintance of his mother. He spoke to the chairman for about 10 min to hear that no recommendation or help was possible. He replied meekly, and the President came to believe that he was faced with a poor, ignorant and, only a little bit, arrogant boy. Serious error of underestimation!

In fact, Pasquale, who had understood the weak point of the man, remained restless only until he saw the lady, who was the examiner for Italian grammar and literature, while talking to the President. Reading the words on their lips, Pasquale could transcribe a poem (bold!) by Catullus. Without this poem having any bearing on the assigned subject, Pasquale put it in the epigraph to his written composition. And in the text, he managed to include a long discussion on clandestine loves. Needless to say, Pasquale brilliantly passed his high school final examination with better grades than his brother. The fact did not go unnoticed: Pasquale perceived a vein of incredulous contempt in the voice of his father, while praising him for this result. With that tone clearly, his father asked himself: *How the hell did he manage to do it?*

University Studies and the First Job

The famous mathematician who examined him in the most important examinations of the early university years had so many vices, and they were all in the public domain, that, in practice, he could not be blackmailed. This was not a true problem for Pasquale: he discovered the weak point of the professor's bigoted assistant. And when the professor arrived late at the examination session (because of the excesses of the night before), he found Pasquale «approved with praise». The professor signed the examination minute but, apparently turning to his assistant, said out loudly: «you gabbed me, but be sure that he must ask somebody else to be his thesis supervisor».

Instead, the famous mathematician personally praised Antonio and suggested immediately that he be his thesis advisor. This circumstance was the end of Antonio's academic career. The famous professor died immediately after the discussion of Antonio's thesis: an event widely foreseen by Pasquale who had understood perfectly the nature of the excesses of that brilliant mathematician. Antonio resigned himself to a position as a high school teacher and lived a quiet life surrounded by the esteem and affection of his students.

Pasquale, instead, asked a mild and depressed professor to be his thesis supervisor.

The depressed professor had been exhausted by six years of collaboration with the greatest mathematician in the country. Pasquale had understood that he would have an easy means of blackmail: the mild professor suffered physically when he had to talk to the students and could not conclude a conversation managing to maintain his point of view. The slightest contradictory remark prostrated him to the point that he preferred to give up rather than continuing the conversation.

Needing an assistant to supervise the students, the mild professor co-opted Pasquale, thinking that, in any case and after six years, he would send him to teach in high school (this was the law in those times!). Without Antonio's help, overloaded with teaching, Pasquale was unable to qualify as professor and was forced to follow his brother in teaching at a high school.

The Return and the Fast Rise

The mild professor had underestimated the network of relationships that Pasquale had woven. His dismissal was seen by the remaining six assistants as the first one of a longer series. They were convinced by a coldly furious (no one has ever seen him raise his voice) Pasquale that all of them would follow his fate. And they rebelled. Even the real mathematicians of the group, those who would have had their career cut short by the return of Pasquale, participated in the rebellion against the mild professor. Who, not knowing how to assert his point of view and not daring to do what he could, namely dismiss everyone and find new assistants, surrendered. He called Pasquale back. However, in an extreme attempt to get rid of him, he proposed him for a professorship in a smaller satellite university.

He thought he had solved the problem. This illusion lasted only for a few years. Pasquale quickly took control of the satellite university, even by helping people of no human or scientific value to get a professorship, and built a powerful network of national alliances. It only took him a few minutes to figure out what to do to reduce everyone to be collaborative. And in a short time, he became the undisputed *dominus* (*i.e. master*) of one of the two factions in which his scientific disciplinary group was divided.

The mild professor discovered that his faculty was «calling back» Pasquale only during the meeting in which the vote took place. The next day he asked for an early retirement. Once back in the prestigious university, Pasquale needed only a few years to take a full control of the situation. He fired all clever mathematicians of the group who did not accept his absolute leadership. He hired obedient, obsequious and prone new assistant professors, who were ready to satisfy immediately all his desires.

He carefully made sure that any original thoughts and innovative initiatives were forbidden, together with any kind of discussion, in his group.

The Empyrean of Science

He had been «called» to be a professor by one of the oldest universities in the world! So, Pasquale could begin to enjoy his power. But, along with the satisfaction related to his exercise of power, he also began to have some problems. He needed to have a long list of publications, for meeting the criteria that we listed before. Actually, using his notes taken from the lessons of the mild professor, supplemented by those that Antonio had left him when he had given up his university career, Pasquale barely managed to publish a textbook. The content is standard, no conceptual leap and no innovation. It was the faithful reworking of the text of the mild professor (Remark that the textbook is dedicated by Pasquale to his father!).

Clearly, the publication of this textbook was not enough. Indeed, Pasquale's further ambitions found another obstacle.

He was very surprised in discovering that the scientific community had accepted a new «fashion»: to evaluate scientists on the basis of their scientific publications! Pasquale has fought all his life against this fashion and has also managed to win several battles. First of all, he managed to obtain a very prestigious professorship, with only three works, published in a local journal, to his credit. Second, he managed to persuade many selection committees to assign professorships to some candidates who had produced only a risible amount of papers of very low scientific quality. Don Pasquale persuaded these committees that his endorsement was more than enough to prove the cultural and scientific quality of the candidates whom he supported.

It is therefore obvious that he never managed to understand the reason why all of a sudden the community had started to consider publications in prestigious scientific journals as a necessary condition for the election to a professorial position.

However, as he quickly understood that it was really necessary to bend to this «strange» situation, he adapted to the «degenerate» modern fashion: he co-opted a «professed» mathematician and gave him the duty to publish scientific papers.

Two things must now be said: Pasquale never spoke, read or wrote in English. Moreover for him, Italian is already a foreign language. His mother tongue is the dialect of his village of origin (this situation represents a strong parallel with Cavour!). It is therefore not possible that he could have written the works that he signed in his long career, using Italian at first and English subsequently. He probably did not even manage to read them!

The professed mathematician managed to convince Don Pasquale (now he had become *dominus*, i.e. Don in ancient Italian) to also hire his own wife. Don Pasquale had not been slow to understand that the professed mathematician was not the true author of the best works that they had both signed: the true author was, indeed, the wife of the professed mathematician! Pretending to comply with the nepotistic demands of his collaborator, he ensured the indirect control of a docile and unpretentious «ghostwriter».

We remark here that the professed mathematician, many years later, divorced from his wife: since the first year after his divorce, he did not produce scientific papers anymore.

At a certain point, Don Pasquale ordered the «ghostwriters» to write a work for the most prestigious journal of his discipline. The task, at that time, was difficult. However, his pressures were enormous: he needed this work. He also wanted to sign it as a single author. After many years, freeing himself from the dominus' yoke, the professed mathematician circulated an informal note on the fundamental error contained in that work. No one cared. And no one wanted to investigate whether the error had been made on purpose or whether the professed mathematician had discovered it only after its publication.

Basing his ambitions on this revolutionary result, Don Pasquale began the assault on the last bastion: the famous academy. He had to persuade the head of the faction against him: and he did it in the best possible way. The opposing leader was a true mathematician: studying all the time, his love affairs could not but be kindled in the academic milieu. Don Pasquale, without any explicit request, worked to have the mistress of the opposing leader elected professor in a prestigious university. The adverse leader, being «a man of the world», opened the doors of the academy to him.

Editor in Kiev

The social successes of Don Pasquale should not make believe that he could play any positive role for the advancement of science, to give any evolutionary advantage to his family, group of humans or any kind of larger social organisation. Evolutionary speaking Don Pasquale's personality is a kind of «peacock tail».

Darwinian theory, in short, claims that the peacock tail is a kind of extreme result of sexual selection. The females started selecting as fathers for their children those peacocks having a heavier and more coloured tail, decorated with mesmerising eyes. This tail became to be the proof of the sexual fitness of its carrier. The evolutionary message sent to the females is the following: the peacock with a big tail is so healthy that it can afford carrying such a useless big tail, notwithstanding its relatively great weight and the related dangers due to the presence of predators. Females started preferring bigger and bigger tails, more and more mesmerising. As a consequence, without any evolutionary advantage, the peacocks were obliged to a race towards bigger tails, even if, in doing so, they are endangering the survival of their species. A mathematical proof of the presence of such evolutions «towards the non-convenient» has been attained and accepted in the literature: such a phenomenon is real, observed (already by Darwin) and now fully understood theoretically.

Don Pasquale is the extreme, paroxysmal manifestation of a sexual selection that prefers more and more cynical and opportunistic alpha males as leaders for primate groups.

In reality, the unfair behaviour of a leader is not convenient in the long run (i.e. in a strategical sense) but it is extremely useful in the short run (i.e. in a tactical sense). These concepts are very well studied in mathematical economy: tactical behaviour favours day-by-day management of situations, and a good leader must have the capacities needed for helping his group survive in the short run. However, a leader who has not a strategic view may help his group to survive day after day but may lead it towards a strategic disaster: especially if he is not lucky as Cavour was!

Don Pasquale (again in this, he seems very similar to the personality of Cavour as described by Brofferio) has no cultural standing.

He cannot speak Italian: he always uses a «rotten» or «broken» Italian language. Indeed, his mother language is the dialect of the poorest peasants in his birth village. He is absolutely incapable to speak any foreign language: once at the beginning of a conference, he read an opening address (written by one of his pupils in English), which he had learnt by heart. Nobody could understand the meaning of a single sentence: but everybody congratulated him. He claims to be a mathematician, but he cannot prove the simplest mathematical theorem without learning by heart the demonstration in some textbooks. He did not manage to lead a strong academic school, as he selected always the worse candidates in any selective procedure which he «governed». Actually, whenever he had a choice between two candidates, he always chose the worse, scientifically speaking, and the most prone. He destroyed a long-lasting academic tradition which assured the teaching of his discipline in any «hard sciences» and «high technology» university faculties. He stopped any scientific velleity of all his «voluntary slaves» (see later sections of this essay) towards any higher scientific activity.

It is infamous his answer when one of his Ph.D. students, having got an invitation to spend six months at MIT, dared to demand his permission to accept. The answer was: «What can you learn at MIT which I cannot teach you better?».

However to understand how embarrassing he actually is, one has to report an event, which has some hundreds learned witnesses, i.e. nearly all the members of his prestigious academy. By presenting to them a speaker in a formal session, he, instead of reading from his notes «he is editor-in-chief of the Journal …», boldly said « he is editor in KIEV of the Journal …». A cold silence fell in the audience. Then, the President of the academy decided to go ahead and joined the speaker in congratulating Don Pasquale for the brilliant introduction.

The shameful ignorance of Don Pasquale is the drawback of his cynical, and therefore extremely effective, conduct in academic policy.

This conduct is so convenient for his voluntary slaves that they oblige, with their prone behaviour, the whole society to pay the terrible consequences of that ignorance.

A Long Sunset

Don Pasquale had finally become a true school master. He has had many students. The most active among them has as his main academic title to have brought coffee during the interval of lessons to Don Pasquale for a long period, say at least twenty years. The second most active is his silent partner in a real estate company.

In presiding over a selection committee that was bound to elect some of his students as professors, Don Pasquale, at one point, stopped a mild discussion between the committee members sharply.

The committee members dared to have an opinion!

They wanted to open, only to open!, the envelopes containing the dossiers presented by the candidates. For God's sake: they did not want to read these dossiers! They only wanted to open the packages so as not to risk formal appeals by excluded candidates.

Don Pasquale with an (only slightly!) altered voice went personally to open the envelopes, returned them to their place, i.e. in the closet where they were kept by the administration, and said a phrase remained infamous: «We do not want to be influenced by the publications of the candidates: isn't it?!?».

His silent partner in the estate company was elected professor.

To become a true school master means, in the particular meaning given to the expression by Don Pasquale, not to give space to scientists who are not obedient to the political line chosen by the dictator. Don Pasquale managed to slow down, prevent and make difficult the career of anyone who opposed him or who simply did not satisfy quickly enough his demands. He managed, in one case, to «settle his scores» twenty-five years after the outrage received, even though he was not part of the committee that cut off the career of the disobedient.

Some disobedient subjects dared to ask him to be forgiven. Sometimes, when his sixth sense told him that repentance was sincere, Don Pasquale readmitted them to his court, accepting to give them new orders. For the other opponents, instead, he had no mercy. He persecuted them until he came to crush the careers of their students and the students of their students.

In the long decline of his life, he had the satisfaction of being able to co-opt new members in the prestigious academy. Of course, these new members' scientific value absolutely does not exceed that of Don Pasquale.

The most infamous among these new academy members is the nephew of a Seigneur of a feudal fief who had wanted to add to the family properties also the succession rights on a university chair and an academic fellowship.

Why the Seigneur did not help his own son to have such a prestigious career? This son was mentally disabled: while his father did manage easily to let him get a professorship, even if in a second class university, he could not, unfortunately, assure for the disabled son a fellowship in the academy. Therefore, he was obliged to choose as his successor the son of his sister. The most remarkable scientific contribution achieved by this nephew consists in a famous *lectio magistralis*. Having being invited to give the most important lecture in a conference in honour of a famous scientist, he stopped the lecture for 10 min to answer to a cellular phone call from his (second) wife. It was urgent: she needed help to find a plumber!

Don Pasquale's father is dead, but he must now keep proving to his first son how good he is. In the hope of being able to break the tradition, this son did not give his own son the name Pasquale. He would have wanted that no «Pasquale» be born again.

Unfortunately, human DNA is the bearer of the dictator's gene. If not the genetic grandson of this specific Don, surely some other child is destined to continue the exploits of Don Pasquale.

The Wife of Don Pasquale

When we described the reproductive strategy of Genghis Khan, we made an apparently trivial looking observation. Only males can have an enormous direct progeny. The Persian chronicles attest to more than two thousands the number of children of Genghis Khan. Obviously, a woman cannot have the same number of children. Although one may therefore think that sexual predator behaviour has an evolutionary meaning only when practiced by males, what is described in the literature of Catherine the Great's appetites definitely contradicts this statement. Thinking about it more carefully, even the claim that only males can have a huge immediate progeny is unfounded. In fact, as we have already observed, also the mother of Genghis Khan has had a huge immediate progeny! What if the evolutionary strategy of Genghis Khan's mother, aimed at having the greatest possible offspring, had consisted in having a son, who was a great leader and greater sexual predator?

Such an apparent paradox can be discussed, in a slightly different context, even when talking about Don Pasquale's wife.

Perhaps, Don Pasquale's irresistible push to power was generated as a form of compensatory neurosis of his wife?

Their First Meeting

The young Artemisia, reading the stories of Greek mythology, always wanted to believe that her father had chosen her name because of a precise awareness. Artemisia is the goddess of female initiations and is, also, the protector of the secret rites of alliances among women. Artemisia was convinced, from her earliest age, that her father had sensed, just at her birth, what the special gift of her daughter was: to know how to hide from all the other human beings her true thoughts, her sincere preferences and her real intentions.

Artemisia could make anyone believe everything.

Her father had fun, all his life, having beautiful stories told by his daughter. All invented, but, all so likely! The lovely Daddy always believed that all the words said by his daughter be true, even when the words she had said before were in clear contradiction with those, which she would have said later.

When Artemisia met Don Pasquale, she also understood that this was the man of her life. Don Pasquale was the first man who had managed to stand up to her. And he will remain the only one for the rest of her life. Don Pasquale will never be able to read in Artemisia's mind, but Artemisia never managed to make Don Pasquale believe something that was not true.

Never? No! Except in one case.

However, this could happen only with the powerful help of a person who could gather the joined talents of the two parents: their second child. That son who, with the help of his mother, managed to make his father believe that he loved him very much, when in reality he was completely in agreement with the very negative opinion that his older brother had of their father.

This second son was, in fact, also a great reader of the mind of others: however, well guided by his wise mother and unlike his father, he began a career in which his talents soon made him very rich. On the other hand, who can stop an intelligent man, who has a powerful family behind him, when this man quickly understands the intentions of his interlocutor and manages to make everybody believe whatever he wants?

Perhaps, the real reason for the indissoluble union between Don Pasquale and his wife is to be found in the powerful drive of natural selection, which absolutely wanted to produce this second child of theirs.

A Winning Team

Donna Artemisia was the vestal of her husband's scientific fame. Her incessant action in support of Don Pasquale's choices has had unparalleled results. Don Pasquale understood the aspirations and true intentions of his interlocutors. With «Donna» (i.e. lady) Artemisia, they decided, as good dictators, who must do what and when. Finally, Donna Artemisia took charge of persuading the pawns in their game to perform the moves prescribed to them. In this refined game, many mathematicians of great value found caught up, without understanding what was happening until when it was too late.

These «great men» initially, and with some arrogant presumption, believed they could easily control two minds that appeared, in their eyes, so simple! Although she has less difficulty with Italian language than her husband, her rather provincial and sometimes clumsy eloquence does not reveal her great depth of thought. There is somebody, among the men who have been deceived by Artemisia, who suspects that also her apparent speech difficulties are part of her strategy: indeed, there is somebody else who heard Artemisia talking in a very elegant way to his second son. There are two possible interpretations of this behaviour: maybe she does not want to hurt her husband by explicitly showing him his limits. Or simply, she wanted to deceive everybody, except her second son. Or maybe, both reasons motivated her.

All the best strategies put into practice by this winning couple have always been elaborated by Artemisia.

Of course, without Don Pasquale's perception of the motivations of their interlocutors, Donna Artemisia could not have elaborated anything. However, the elaboration and realisation of strategies have always been exclusively hers. Indeed, it was her sweet or forceful persuasive action which managed to deceive so many mathematical geniuses and in such an effective way. Donna Artemisia has always perfectly calibrated her words. They are accommodating towards the powerful and famous academicians: sometimes, she is not only persuasive but also docile. Powerful academicians have always been misled with education and courtesy, at least in the early stages of their interaction with our winning team. Clearly even the most famous scientist, after having accepted some personal favours, is then treated with impatient condescendence and must learn to please Donna Artemisia quickly. If, on the other hand, the interlocutor, whose behaviour and decisions have to be dictated, is a scholar of subordinate rank, then Artemisia can show an unparalleled brutality in giving dry and hasty orders. In a group where all decisions must be shared by collegial bodies, the ability to suborn the individuals who participate in the formation of decisions has been the trump card of Don Pasquale.

A Process of Transference

Artemisia's father was a man of great intelligence who, by birth and by the decision of fate, was unable to obtain all the awards, which his daughter was persuaded that he deserved.

Artemisia wanted to take revenge on fate: she wanted this revenge very strongly. The dowry that her father lacked was that sort of clairvoyance that Don Pasquale had allowed her to exploit. Artemisia knew well that Don Pasquale did not even remotely have the great intellectual gifts of her father. She was also aware that scientific talents are not so important to have success in the academic environment in which Don Pasquale had decided, for the reasons described above, to want to excel.

Artemisia wanted to create simply the appearance of a great scientist from nothing, with the intellectual raw material (not really exceptional indeed) that she had managed to find.

Surely, in fact, the qualities of intuition of the thought of other persons shown by Don Pasquale are certainly not the qualities, which, in principle, are to be required to become a scientist. Don Pasquale has not been, even in a single moment of his life, a scientist. However, his ability to read in the thought of (wo)men, together with his wife's persuasive talents, was sufficient to create the appearance of a great academician.

The descendants of Artemisia (her second son in particular) would represent, eventually, her revenge against that adverse fate, which had been raging on her father. And note that while the baptismal name of his first son was that of his paternal grandfather (circumstance which, almost as a sign of destiny, imposed on that son to share the opinions of this grandfather), the baptismal name of the second son was that of his maternal grandfather. And as we have already said, it was precisely this son who brought the glory of the family to the maximum possible extent.

Artemisia did not seek glory for herself.

Her social background, the tradition and the habits of the world in which she was born, all these circumstances imposed that a woman could not be, in the eyes of the world, a dictator (obviously, in the sense of Arrow!). In that world, however, it is licit, and even tradition, for a woman of value to impose herself through her husband and her children. Artemisia succeeded in this feat perfectly, proving that she is a successful big-woman. The power acquired through her husband, the academic prestige he achieved, the network of alliances he managed to create were all aimed at launching her son into a world whose existence both she and Don Pasquale had never even managed to imagine.

The two of them were always treated as parvenus in that world. Their son did not. Their son could fully dominate that word!

Patroclus Petrons: The Puppet of the Crypto-dictator

It is very difficult to be a dictator. The group, which is choosing an individual as a dictator, demands to this individual to be capable, reliable, altruistic and empathetic, or at least, it demands to him to pretend to have such talents.

De Waal has studied the onset of empathy in primates as one of the first signs of superior intelligence, which has been developed. Empathy and intelligence are very often inseparable twins. Individuals who do not have the needed gifts to be accepted as a dictator do not manage to maintain their power. Therefore, if an ambitious individual seeks for power, without having the personal talents needed to maintain the role of dictator, then he can try to exert this role by hiding himself behind a puppet. The use of puppets is often unavoidable when the candidate dictator is explicitly and clearly unsuited for this role.

We can call this kind of leader a crypto-dictator.

As puppets are disposable, the crypto-dictators can change them when the group is tired of such an incapable dictator or when the puppet decides to try to get rid of his puppet master. Very often, the group believes to have managed to get rid of a dictator, which is not suitable to its well-being, but in the darkness the crypto-dictator operates to keep his true power intact by finding immediately a new, more presentable and docile, puppet.

The rational study of the dynamics of social groups must necessarily take into account this phenomenon: the systems of warranties, the check and balance of powers and the evaluation *ex post* of a magistrate and his actions, all become vain if the change of the officer which results after his replacement as a leader is only apparent and the same person remains *de facto* in charge.

When the puppet master is very clever, then he can afford to have a rather clever puppet. In this case, he wants to play this manipulative game only for avoiding the rules, which, usually and wisely, limit the duration of the dictator role. In this case, the damage, which is imposed on the social group, is not too big.

A clever dictator manages to keep his power without harming his group.

Indeed, he knows that if his action is not altruistic enough then soon or later either he will lose his power or he will manage to destroy his group.

In reality, very often, puppet masters decide to have this role because they are aware of their limits. They know that they are not clever enough to be good leaders. In this case, they must choose puppets who are less clever: i.e. definitively stupid. They manipulate their puppets by flattening them, by menacing them or by rewarding their fidelity with indecent favours. This social mechanism is very detrimental for the survival of the unfortunate group falling under the influence of a crypto-dictator. Puppet masters are so harmful that they can manage to make appear Don Pasquale dictatorship as preferable.

How can a constitutional mechanism be conceived to diminish or to cancel the possibility of the existence of a ghost dictator and a series of his puppet dictators?

We are afraid that until voluntary servitude and stupidity will be diffused in the humankind the phenomena described here cannot be easily uprooted. However, we will try to put forward some proposals to attain this uprooting after having examined some aspects of the here considered phenomenology.

The Gifts Needed for Getting in Power Are Not Always Those Needed for Exerting It in the Interest of the Group.

Since his first childhood, **Caesar Sacristy** had suffered for his height rather below the average. This was a clear sign of his ancestry: being of Southern Italian origin, he had not many chances to be tall, to be blonde and with blue eyes. His wit was quick, and his capacity in manipulating human beings was also inherited from his far Greek origin. He chose immediately a schoolmate who could help him to manipulate his voluntary servants: already during his first elementary school class. Caesar was aware of his intellectual limits: he never managed to speak correctly in English, and his elocution was never elegant. His mother language was a mixture of the original South Italian dialect spoken by his father and mother and the rudimental English spoken in his suburb. He never managed to polish his way of speaking, above all because he never tried to do so, finding more efficient to exploit somebody for writing his homework and later his papers. He managed so effectively to dominate his first victim that the poor child, also after having become adult, never managed to free himself from his voluntary slavery. This sociological phenomenon is very frequent: we will limit ourselves here only to observe its effects.

The preferred victim's name was Octavius.

Octavius followed the orders of Caesar until the last moment of his life: many observers recognised, in different moments of Octavius life, the symptoms of the Stockholm syndrome. Somebody said, as a joke, that Octavius demanded to Caesar the permission to die, when his disease became too painful.

The more evident it was that Caesar was exploiting Octavius (in any possible way) the stronger was the obedience shown by Octavius to the desires of Caesar. Octavius helped in every possible way Caesar and, as he was diligent enough, he managed to construct a scholarly career to both of them. Of course, in academia, Octavius remained always one step behind Cesar: he never dared even to think that he could surpass his «master».

Caesar is capable to understand how to manipulate the expression of the choice of any social group.

He manages to persuade everybody to follow his orders. The persuasive process is very simple: Caesar manages to obtain obedience by promising to every victim what this victim desires. He leaves open in his discourse the possibility that this desire could be not satisfied: but he manages to make the promise realistic enough.

In his life, Caesar has promised the same position, the same grant, the same elective role on the average to five persons. One of them has always got what had been promised.

The gain of Caesar is that he is bribed by five persons and pays one favour back, only. In this way, he could gather an enormous personal patrimony without really losing his reputation.

A diabolic behaviour indeed!

Satisfying one of his voluntary servants, he could always prove that his promises were not unrealistic. He could always prove to the remaining, disappointed, four persons that their failure had been their own fault. They had been not enough obedient, and therefore it was right that somebody else had deserved the promised prizes.

On the other hand, Caesar was totally unsuitable to be a leader in the scholarly group, which he had decided to dominate.

Even with the help of Octavius, the research papers, which he signed, were not very meaningful. Also, his political skills were not very constructive. He has applied his whole life the principle of Roman, then Venetian and then British Empires.

Divide et Impera.

He has systematically spent his life by spreading fake news, completely invented gossips, imaginary plots and attributing to everybody completely false intentions. He became so expert in inventing this fake reality that, in all groups were he managed to get leadership, terrible quarrels, endless enmity and bitter animosity spoiled the interpersonal relationships. He has been always the leader in very fragmented groups torn apart by extremely bitter intestine wars.

Clearly, his capacities are very effective to get power, but dramatically harmful for the well-being of the social groups having the misfortune to have selected him as their leader. For this reason, the most influential members of the group he dominates decided systematically to remove him from power, and for this reason he learned how to become a puppet master.

The Long Series of Puppets

The constitution of Western societies is based on the wise principle of the alternation of dictators, in the check and balance of powers and in the plurality of centres of powers.

Caesar's family had a tradition coming from his cultural roots. The lord of the castle dominating his original village could survive the unification to the Italian Kingdom by bribery. He found an effective way for continuing to exploit his serfs, and with the obtained richness he could buy his Senator's chair in Rome, his son's general position, his son-in-law's degree in medicine and so on. Caesar wanted to parallel the deeds of his Seigneur. He understood immediately that he could dominate his group of schoolmates with his psychological manipulation. However, he knew that in order to reach the highest peaks of powers he needed to control a strong economical power.

Therefore, he decided that it was necessary to marry a very rich woman and to use her economical strength to establish strong alliances.

Actually, he became the owner of several companies, all of them investing in the technical field in which he had obtained a professorship in a prestigious Ivy League university. Using his academic prestige, he made these companies very wealthy and he could start associating persons to his power system in a very simple way: giving them some shares of his companies, or paying them, based on fictitious consulting contracts, a part of the money gathered with the revenues of his consulting activities, all obtained thanks to his political friends. Soon, he understood that the family links that he had inherited from his father-in-law were not enough. He managed to be accepted in a network of professionals gathering every Saturday morning, in exclusive clubs, to establish the needed links for being helped in the most important economical activities of his companies.

He managed to be elected as dean, then as provost and finally chancellor of his university.

Unfortunately, his leadership was very weak, his capacity of choosing clever scientists absolutely null and his honesty clearly inexistent. Therefore, a group of professors started a liberation war against him, by challenging his power publicly. His biggest mistake was to sell a big land property, owned by the university, to a private enterprise, at a very low price. This caused a big scandal, and even if he managed to avoid any criminal prosecution, he was obliged to resign and even to retire. However, his economical, political and social powers were left intact. His group of friends was always there to help him in every situation. And, most important circumstance, he could keep manoeuvring Octavius. The poor victim now was free from the obligation to write papers, which had to be signed by Caesar, as Caesar was now retired. Therefore, Octavius could dedicate full time to the job of participating in all committees to which, because of his retirement, Caesar could not participate. When talking about his retirement, Caesar seemed a lion pacing and roaring in its cage: he felt as somebody to whom it had been subtracted a property owned by many generations of ancestors. He felt to have a kind of divine right on his academic status and offices. However, for a while, he accepted to exert his power via Octavius. Unfortunately, also Octavius was close to his retirement.

Then, Caesar found a young assistant professor, who had all the characteristics needed to be the perfect puppet in his hands.

Indeed, the assistant professor we are talking about has no human or scientific gift: he is unlearned, decidedly ignorant, incapable to write two sentences having a precise meaning or to pronounce two sentences without infringing grammar or syntax, absolutely incapable to lecture in a reasonable way, totally lacking the ability to conceive what original research is.

For all these reasons, Caesar choose Patroclus Petrons as his puppet.

He sent him to study a few years in a satellite university, where a clever scientist was trying to establish a good scientific school. The clever scientist, too involved in his scientific challenges and being, by nature, rather naïve, accepted to support Patroclus in his first scientific steps. However after many efforts to introduce him to scientific research, he gave up: Patroclus was absolutely incapable even to perform simple calculations under the direct guidance of a supervisor. However, Caesar had been very clever in manipulating the clever scientist: he deceived him by promising a strong support for his school as a compensation for the help given to the career of Patroclus. Instead, as soon as Patroclus was ready (i.e. had enough papers and enough citations), Caesar called back him in the biggest university and forgot all promises. Patroclus became quickly full professor and then elected dean. In his office, Patroclus asked to place a second desk: the biggest one was where Caesar was sitting and giving orders, and the second one was for him, acting as a secretary of the true boss.

The network of power supporting Caesar was happy: the interests of its members could be safeguarded as if nothing had happened. Everything seemed to have been settled until a woman caused some extra troubles: she was ambitious and wanted to become professor. Being Caesar not tall enough and decidedly too old, she decided to become the mistress of Patroclus. The alliance of Caesar and Patroclus, exactly as happened to the alliance between Yeroen and Nikkie against Luit (see the corresponding sections in this book), was seriously harmed by this choice. First of all, Caesar is intrinsically envious of every success which other people may manage to attain.

How it was possible that his puppet had a nice mistress and that this mistress did not prefer the true power holder to a simple puppet?

Second, Caesar was worried: this mistress could have started soon to give contradicting orders to his puppet!

He could not avoid to start a campaign against his own creature, denigrating him exactly as he had done for everybody else in the past, including the poor and faithful Octavius. Patroclus was quickly burnt, he became academically irrelevant, the ambitious lady was obliged to have a love affair with Caesar, and a second puppet was found by the true power holder.

The academic group ruled by Caesar did not waste his time to memorise the name of this second puppet.

Neither it did this with the third, the fourth and so on. Indeed, the eponymous hero in this story is Patroclus: everybody in that faculty called the current simulacrum of dean with the name of the first puppet. For everybody, during the reign of Caesar, the apparent dean has been always referred to as «Patroclus». This phenomenon is well known in history: all Roman emperors called themselves, and were called by everybody, Augustus, i.e. with the name of the first emperor.

Only the physical death of Caesar managed to change this power configuration, leading to the so much waited change in the disastrous management of that unlucky faculty.

When finally Caesar died, and only then, everybody dared to use openly the nickname that everybody had used secretly during all his years of power: «Godfather».

Big-men Are the First Observed Form of Crypto-dictators.

Big-men are defined as the most influential individuals in tribal societies, following the analysis due to Sahlins (1963), which was based, at first, on the observation of the social structure of Melanesian and Polynesian tribes.

Tribes are those societies, which can be defined as social groups in which no formal rules for regulating the social life are established.

Of course, the wisest and eldest men in these groups refer to tradition, religion and some form of ethics for enforcing some «socially imposed choices» on the other tribe members. We are aware of the debate among anthropologists concerning the concept of tribe, and we are also aware of the misuse of the concept, which characterised the ideology of scholars during the colonialist period.

We refrain from any discussion about this misuse, as here we simply use the word «tribe» in the sense of «first kind of aggregation of human individuals», i.e. of «intermediate step of organised societies preceding, in the evolution process, chiefdoms and States».

Roughly speaking, we like to follow the system of classification proposed by Elman Service (1975), who tried to characterise the observed social organisation in all human cultures. He analysed and described the presence of social inequality, the way in which individuals' interactions and conflicts are ruled in different societies and the structures which anticipated the occurrence of the State.

Elman Service's system of classification consists of four categories of social structures:

(i)   Hunter-gatherer bands, which appear as fundamentally egalitarian, with a pecking order established by specific ritual «duels» (remark that this kind of society is observed also in chimpanzees and bonobos troops),

(ii) Tribal societies showing some forms of organisation, based on social rank and prestige (in these societies, one can observe big-men or big-women),

(iii) Complex tribal societies, with stratified social classes, led by chieftains (these are the so-called chiefdoms),

(iv) Civilisations, with complex social hierarchies and organised, institutional governments (at this organisation level, one can use the concept of State, in the sense given to it by modern jurisprudence).

In general, Melanesian and Polynesian tribes apparently base their social life on egalitarian principles and on the refuse of violence: however, this appearance must be reconsidered carefully. Indeed, some forms of cannibalism have been described also in apparently peaceful tribes and the interpretation of reasons for such a social phenomenon is not easy. We believe that this form of cannibalism is an indicator of the ubiquitous presence of the innate tendency to violence, which humans share with chimpanzees and gorillas. Remark, on the other hand, that, like bonobos, many Melanesian societies are organised in a kind of female-dominated society (as matriarchal, or matrilineal or matrilocal or matrifocal societies). While we will not discuss here the structure of all these kinds of societies, as described and debated by ethnologists, we want to note that if in social groups a relevant share of power is in the hands of females then the incidence of violence for solving social conflicts is much lower. However, also in such low violence, essentially peaceful and apparently anarchist societies dictators have been observed. In these societies, there is essentially no other social structure, except pecking order.

One comment is probably needed here: one should not believe that the use of violence is written indelibly in the DNA of human beings. Indeed, it has been observed that when one tribe of Maori people did move to Chatham Islands, because of the change of the environment where it had to live, it was transformed into a very peaceful tribe: notwithstanding the fact that Maori are well known for their military traditions and habits.

The concept of big-man and big-woman has been formulated, to distinguish the kind of dictators observed in tribes from chiefs, defined as dictators whose role is formally recognised.

Big-men and big-women are persons who do not have formal authority of any kind (they do not control, for instance, material possessions or do not enjoy recognised inheritance of rights or do not have a recognised religious role), but manage to maintain recognition through skilled persuasion and wisdom. One should not give any, a priori, positive or negative meaning to the adjective «skilled» and the noun «wisdom», here. For instance, skill could be the capable exercise of violence, backbiting, slander and manipulation, and wisdom could be a true knowledge of facts, which are important for the survival of the group, but also a deceptive capacity of imposing superstitious beliefs.

Every big-(wo)man must have a relatively large group of followers.

Their role is attained and kept by providing followers with protection and with economic or social or psychological assistance. The rewards obtained in change of their protection are the social support received by «protected» individuals: this support is used by big-(wo)men to increase their status and influence.

There are big-(wo)men who are using the accumulated capital of reputation and social status to improve the well-being of their group.

There are big-(wo)men whose final objective is their own exclusive personal interest only: obviously, the groups in which this last kind of dictators arise are more fragile.

Big-(wo)men have no formal recognition of their role, which is simply based on their reputation.

Many examples of a sudden change of the status of big-(wo)men have been observed. For this reason, we believe that they can be regarded as the prototype of crypto-dictators. The main form of power in the hands of big-(wo)men is their capacity of persuading the other group members to follow their desires, without exerting any formal and explicit influence or without giving any explicit order.

Don Pasquale, his wife and Caesar, whose ventures have been described before, are examples of big-men or big-women in some specific tribes of academicians.

There are, however, many big-men and big-women who altruistically worked for improving the quality of life of their group. Everybody did meet in his life this kind of person. The most miserable among us did exclusively exploit them and their help. However many times, the cooperative behaviour induced by these altruistic big-men and big-women induces, by emulation, very positive and constructive behaviour in their group, possibly leading it to great endeavours.

How to Avoid the Presence of Crypto-dictators in a Social Group

The crypto-dictators who have bad intentions may manage to infect even a well-established legal social system, leading it to the total collapse.

On the other hand, also crypto-dictators inspired by good intentions may cause a violent reaction of their opponents, as they can be perceived as tyrants. Of this last phenomenon, many historical examples can be given.

One of the most famous examples concerns Lorenzo il Magnifico, the *de facto* Seigneur (i.e. crypto-dictator) of Florence in the flourishing period of Renaissance. Florence, like many Italian *Comuni* in that period, was a democratic State in which an institutional system of government had been established. However, the enormous economical power of Medici bank and the ambition of Lorenzo (*and of his mother*) to lead a process of political unification of Italy produced many historical events. The reaction of Pazzi family, competitors of Medici in both banking and political arenas, is infamous: being supported by the Pope, they tried to kill all males of Medici family, during a religious service. The substantial failure of Pazzi conspiracy allowed the Medici family to become, eventually, the formal rulers of an Italian State (Tuscany) for a long period.

However, the most famous example, in Western culture, of a crypto-dictator who became, passing through several episodes of brutal violence, the initiator of a long sequence of emperors is Julius Caesar. He was elected many times to the office of consul, and only after having gathered an enormous patrimony (whose heir was Octavian Augustus), and a long list of military successes, he managed to have recognised his dictatorship role formally.

There is only a method, found during the history of humankind, to avoid, or at least to limit, the presence of crypto-dictators in organised legal States. This method consists in establishing strict control of every form of power by formally enforcing a system of check and balances.

The Constitution of USA gives a very successful example of such a legal system. We do not want to say that in USA there were not crypto-dictators who could influence the federal policy. We simply observe that no such dictator did manage to take absolute power, as instead it was possible to Mussolini or Hitler, who did manage to exploit the flaws of not-so-perfect constitutional systems.

We believe that the biggest weak point of Western academic institutional systems consists in the absence of a control system of the power that some crypto-dictators inevitably manage to gather.

Caesar Sacristy did operate in a social group that is closer to a tribe than to a State. Even if already Frederic II Hohestaufen, when founding his University in Naples, had understood the need of fixing strict rules for the functioning of academic bodies, it is true that, still nowadays, the evolution of academic groups is blocked, in the best possible situations, to the stage of a chiefdom.

In very few instances (i.e. in the French university system), a very primitive legal system has been imposed on the self-governance of the academic body. In Italy, a good law for ruling the academic life (the DPR. 382/1980) has been deprived of its better features by reforms, which allowed again the proliferation of academics tribes and chiefdoms.

A Tentative Explanation of Some of Described Phenomena Based on the Concept of Nash Equilibrium

Here, we try to present a preliminary theoretical explanation of the huge amount of phenomenological evidence, which was presented in the previous pages. We are aware of the fact that many other pages, like those which we wrote about the Theorem of Arrow in the second appendix, may be needed to explain in detail those mathematical ideas which we will use only intuitively here. We hope to have the possibility to produce these more detailed pages for future work. For the moment, we try to simply transmit their «flavour» to the readers, hoping to be clear enough to persuade them about the validity of our reasoning.

The focus of our discussion now is the following question:

Given that one (or more) dictator(s) is(are) always logically needed in a social group, which is theory describing the process of selection of the dominating dictator(s)?

The phenomenology, which we have described before, will give us some indications about the model to be developed.

Game Theory: A Powerful Conceptual Tool

In particular, the previously presented phenomenology of leadership and dictatorship shows how important is the selection process that produces the social choice of a dictator.

Therefore, the question just raised is undoubtedly of great relevance. We believe that the needed conceptual tools to solve it are not far from being available. Indeed, the mathematics of twentieth century developed the beautiful «Theory of Games» whose founders are Oskar Morgenstern and John von Neumann. The theory of games studies rigorously, among many others, the problem of determining the optimal strategies in any cooperative or non-cooperative game and the problem of finding the configurations of equilibrium in the competition among different players.

© Springer Nature Singapore Pte Ltd. 2019
F. dell'Isola, *Big-(Wo)men, Tyrants, Chiefs, Dictators, Emperors and Presidents*,
https://doi.org/10.1007/978-981-13-9479-9_7

In the particular context that we are considering here: the game is the game for getting leadership in a social group.

The rules of the game are determined by the laws that are really enforced and/or by the possibility of the players to use different forms of violence, in absence of laws or in absence of the enforcement of laws.

In any case, a specific game is determined by: its players, their current resources and the rules that are enforced to limit the actions the players.

For instance, in the case of the competition for power in Italy, before unification, the players were at least: all Italian independent states, British Empire which dominated Malta and had great economical interests in Sicily, France which controlled Corsica and was allied or protector of many Italian States, Austro-Hungarian Empire, which controlled the North-East of the Peninsula. Formally, these players declared to accept some basic international laws, but in practice they used every means to try to dominate the others, including war, bribery, guerrilla, economical exploitation, financial blackmailing and everything else that their leaders could conceive as a useful tool. Consider that Cavour, in order to persuade Napoleon III to help Piedmont against Austro-Hungarian Empire send the Countess of Castiglione in Paris, with the explicit task of sexually subjugating him: she was so beautiful and smart that she managed to fulfil perfectly her task.

Another case is the game played until the last instant of his life by Don Pasquale. The set of players is the set of all scientists of his scientific discipline; the rules are the few rules which the State enforces on the academia for allowing the access to the resources made available for hiring new scientists and financing new researches. Don Pasquale had many opponents: he always managed to find the correct tactical moves to persuade all the other players to act in the specific way that favoured his plots.

To make our considerations more precise, we need to specify better what is a game.

A game is specified when one has fixed:

(i)   the set of players,
(ii) the parameters specifying the configuration of each player at every instant,
(iii) the set of admissible «moves» which can be performed by each player,
(iv) the set of rules determining the changes of the state of the players after every «confrontation», which results from the moves of the players.

Let us try to give some details for explaining what we mean.

Each player has, in every instant, a «capital» of resources available for trying to win the game. The set of parameters describing this capital specifies his configuration. He can decide to invest part of this capital in the subsequent move of the game. Each player has to choose how much of his capital he wants to invest in every move. The rules determine the effects of all-players-chosen moves, by determining how the output of the current moves changes the configuration of each player. A set of all-players-chosen moves can be positive for one player, increasing his capital, and negative for another player, decreasing his capital. After every move, the configuration of all players has to be changed as a result of all the players' choices. The values of the parameters determining the configurations of the players establish who, eventually, is the winner of the game. Of course, in complex games, the configurations of the players may change continuously with alternant effects of each move on the dominant position of one or another player. It may happen that a move, which gives an immediate apparent positive effect for a player, will eventually lead to his total defeat.

Those who know about chess games will recognise many similarities in the language and concepts used here and, indeed, many schools forming diplomats or politicians used chess game for training the apprentices. A remark is now needed: chess game is completely understood by means of modern mathematics and computer science. As a consequence, no human can win a chess game against a computer.

Instead, the game for power is, for the moment, too complex for having been fully modelled mathematically: however our Archimedean (and Epicurean) philosophy persuades us that we simply need to wait some (longer or shorter) time, and also this game will be formalised and completely described mathematically.

We are aware that our point of view is not completely shared by all mathematical physicists. The debate about this point is beautifully resumed in Arnol'd (1998):

> The mathematical technique of modelling consists in [...] speaking about your deductive model as if it coincided with reality. The fact that this path, which is obviously incorrect from the point of view of natural science, often leads to useful results in physics is called "the inconceivable effectiveness of mathematics in the natural sciences" (or "the Wigner principle"). Here we can add a remark by I. M. Gel'fand: there exists another phenomenon comparable in its inconceivability with the inconceivable effectiveness of mathematics in physics noted by Wigner, and that is the equally inconceivable ineffectiveness of mathematics in biology.

We believe that both Epicurean philosophers and Gel'fand may be right: indeed Gel'fand refers, most likely, to a specific class of mathematical methods: those available in his epoch. Instead, Epicurean are ready to wait for some millennia until some mathematicians are able to develop the right conceptual tools.

(Stable) Equilibrium Configurations

It is clear that evolutionary problems are rather complex to be analysed. A preliminary step in the analysis of considered phenomenology may consist in the determination of the equilibrium configurations for a given game.

What is an equilibrium configuration?

There are different definitions of equilibrium configuration, and these definitions capture different aspects of the involved phenomena. These definitions may be more or less general, and under a certain set of assumptions, which we will try to specify intuitively, are equivalent.

An equilibrium configuration is a configuration in which the system remains unperturbed during its natural evolution. In other words: if the initial configuration is chosen to be an equilibrium configuration, then the natural evolution of the systems will be: remain there!

It is clear that every tyrant or dictator or wise leader aspires to place his social system in an equilibrium configuration.

In few and happy situations, this aspiration was transformed into reality: Augustus managed to establish the **Pax Romana or Pax Augusti**, celebrated by all his apologists; Francisco Franco kept Spain in an equilibrium configuration notwithstanding the raging of the Second World War; the United States of America, after their foundation and the British–American War, were never invaded by foreign powers, and experienced only one civil war; European Union assured 70 years of absence of wars, in a continent which has been torn by incessant wars since the collapse of the Roman Empire.

However, external perturbations may move a system away from its equilibrium configuration.

An effective social system should not only exhibit equilibrium configurations, it should also be stable under external perturbations. Roman Empire did not succeed to manage the crisis ignited by the arrival, at his borders, of the Goths and Alans. The Battle of Adrianople (9 August 378, also known as the Battle of Hadrianopolis) in which Eastern Roman Emperor Valens was killed and his army completely destroyed is often considered the perturbative event which eventually caused the dissolution of Western Roman Empire: remark how the paths of history are sometimes difficult to predict. The defeat of the Eastern Roman Empire, which had not been helped by the Western Roman Empire because of a cynical opportunistic move, did eventually cause the final fall of this last.

It is therefore useful to consider a particular class of equilibrium configurations, i.e. **stable equilibrium configurations.**

These configurations are «resistant» to external perturbations. If an external perturbation displaces the system not too far from a stable equilibrium configuration, then the system remains close to it. Moreover, stable equilibrium configurations are configurations towards which the system must converge «naturally» after its time evolution if it starts from a specific set of initial configurations, because of some «damping» effects.

The set of initial configurations starting from which the system «falls» towards a specific stable equilibrium configuration is called basin of attraction for given equilibrium configuration.

Non-uniqueness of Equilibria

We believe that the historical patterns observed by Giambattista Vico are some «stable» and «recurrent» configurations, as observed in the «game for power» played inside human societies.

The following kind of advanced states have been observed: Kingdoms, Republics and Empires. Rome institutional organisation experienced all these three structures for the State, in the given sequence. Remark that also France did experience during its revolutionary period exactly the same sequence as Rome, while Britain developed a kind of Republican Kingdom, in which it stably blocked, albeit later such a Republican Kingdom formed an Empire, whose Emperor was, *de facto*, a Parliament.

We could imagine that in a more general Mathematical Theory of the State, these three structures for the State are kinds of stable equilibrium configurations. Clearly, the initial state of France, after the collapse of Roman Empire, was in the attraction basin for kingdoms: The Merovingians and then Carolingian dynasties took power, and Monarchs ruled the Franks until the French Revolution. After a Republic without a sufficiently strong constitutional system, usually, Empires are established (see, e.g. Augustus and Napoleon rise after the collapse of the preceding Republics): therefore «weak» Republics must be in the attractive basin for Empires. A society, in which individual freedom and a reasonable well-being are not diffused enough, is subject to Revolutions or abrupt changes of State organisation (see for instance, what happened in India before the end of the British Empire or in Soviet Union before its collapse): we can imagine that some conditions for instability may be developed for a large class of societies.

We are clearly very far from an exhaustive theory, but it seems clear that the ideas presented before have a solid conceptual ground.

The previous considerations suggest that, in general, in social systems there are multiple possible stable equilibria. Accidental and fortuitous events or the genius of great leaders may drive the considered social systems towards one or another among these equilibria. In this context, the Theory of Probability may play a relevant role: it is, indeed, universally accepted that the fate always plays a relevant role in human events. However, if one has a specific structure of a society and the set of stable equilibria is determined, the fate can choose only among the available options!

In the game for power, in general, several possibilities present themselves for the given society. Many of them are stable configurations. However, initial configurations, external perturbations and *the fate*, may drive the system towards one or another of the logical possibilities. In particular also in the game of selection of the dictator, there are, in general, several possible equilibria.

Among the questions, which arise in this context, we focus on the following ones:

There exists one or more equilibrium hierarchical order in the set of players. How it is established? Can we predict it? Is it stable or it may change? Can we conceive a modification of the rules of the game in order to lead the system to determine as his dictator an individual who cares about the group as a whole?

Few among these questions can be solved by means of the results that have already been found by the mathematicians working in the theory of games, and we will try to give an idea of these preliminary results.

Remark here that the questions listed before deal only with the set of equilibrium situations concerning the pecking order in a social group, but the general concepts we will evoke have a much wider range of applicability.

One can consider, for instance, the problem concerning the tendency of some societies to develop revolutions or the reasons for which other societies are too much prone to develop organised criminality. The author would personally like to understand why in the capital of his beloved Nation, i.e. Naples, the respect of laws and rules is so difficult to be enforced.

Of course, these exemplary problems will need a complex and longer analysis, not completely and precisely understood yet.

Strategies in a Game

Once the rules of the game are fixed, each player has left some individual choices, which he can arbitrarily decide in playing his game.

A strategy is the algorithm which a player chooses to determine the sequence of his moves (or game options), among all the possible algorithms, in a game where each moves outcome is determined not only by each player-specific actions but also by the actions of other players.

The concept of «strategy» must NOT be confused with that of «move».

For instance, Don Pasquale strategy consists in understanding what a competitor really wants and then in conceding it, only if in the change he gets a profit which increases his own power. Also, the strategy of Caesar Sacristy consists in promising to more than one candidate the same position, getting favours by all of them, by assigning the position to only one of them. While a move of Don Pasquale consists in giving a certain position to a certain candidate, or a move of Caesar consists in paying a certain consulting service to a certain member of a selection committee for determing the choice of the candidate whom he prefers.

In short, a move is a specific action of a player at some specific stage of a game while a chosen strategy is the chosen complete algorithm for playing the game, which predetermines any move in any possible game situation.

A strategy dictates a priori to a player how he has to act in every possible situation in the game while he is playing. Once a player has chosen a strategy, he can «mechanically» play his game by simply following his predetermined actions sequences.

Finally: **a strategy profile is a possible set of strategies acted by all players.**

A strategy profile fully specifies all possible moves that will be acted in the considered game by all the considered players. Obviously, in a strategy profile, there is one and only one strategy for every player.

Nash Equilibrium

Now, we are ready to discuss the most fundamental concept developed up to now in game theory.

A Nash equilibrium, named after John Forbes Nash Jr. (Nobel Prize in Economy for 1994), is a configuration in the strategies of the players (i.e. a strategy profile) in a non-cooperative game (with two or more than two players) in which

(i)   each player perfectly knows the chosen strategies of all the other players,
(ii) each player assumes that the other players will systematically choose the strategies inflicting to himself the biggest possible harm,
(iii) each player cannot gain anything by changing his strategy, in presence of the choices made by the other players as said in point (ii).

In other words, if each player has fixed his game strategy, and no player has an interest in changing his strategy (assuming that the aim of all the other players is inflicting to him the maximum possible damage) while the other players do not change theirs, then the chosen set of strategies and their corresponding payoffs is called a **Nash equilibrium**.

The big result given by Nash is that:

Every finite game has a Nash equilibrium.

The reader will remember that the thesis of a theorem can be true only when its hypotheses are true. Therefore, the hypotheses of a theorem must be well specified.

What is a finite game?

A finite game is a game with a finite number of players and a finite possible strategy profiles.

Therefore, one must consider that Nash theorem may not be applicable if, for instance, the possible choices for the involved players are infinite.

More sophisticated analysis may be required for demonstrating that a game, after a transitory period of adaptation, always stabilises by «falling» into a Nash Equilibrium.

We believe that a complete understanding of the theory of dynamic games has not been attained. We therefore refrain, here, to delve into this subject. We believe, however, that the last conjecture is rather generally true: Nash equilibria are, for a large class of systems to be determined, the final configuration towards which, finally, the social systems evolve.

Some Tentative Applications of Nash Theorem

It is clear that **only stable equilibria**, in the game for power and also in all other games, **can be observed, as repetitive patterns, in social phenomenology.**

Actually if many external perturbations can move the system far from an equilibrium configuration, it is clear that such a configuration cannot be observed too often.

The idea that Nash equilibria are those configurations towards which social systems are converging is very suggestive and leads to interesting related conjectures. We believe that all observed phenomenology described in this essay must be a form of Nash equilibrium if the observed facts show sufficiently permanent features.

For instance,

the patterns distinguished in historical events by Giambattista Vico most likely refer to some kind of Nash equilibria.
Vico's greatest (and virtually unique) work is the *Scienza Nuova* (1725, New Science). He hoped to introduce, by means of his intellectual efforts, a science that manages to describe human history exactly as Newton did manage to describe the phenomena occurring to moving bodies. His aim was not only to «record» but also to «explain» the historical cycles by which societies «rise and fall». To Vico is attributed the expression «corsi e ricorsi storici» (cycles and counter cycles of history). For him, "the cycle" (corso) or "the counter-cycle" (ricorso) of the «cose umane» (human things) are oscillations of the events in political life, social organisation and order. Events produce jumps back and forth, when following their «metaphysical principle», i.e. (if our interpretation of Vico'e expression is correct), when following the general principles of the theory of societies and States that more suitably describes them. We claim that one of the principles to be used is: a social system tends to reach a Nash Equilibrium.

We find very suggestive to interpret Vico's ideas as follows:

There are many Nash equilibria in social structures and history reduces to a jump from one equilibrium configuration to another or to oscillations around a given equilibrium configuration.

Vico was a very erudite scholar grown in a very formal cultural environment: his presentation therefore relies on a very complex use of words, many of them being neologisms founded on intricate etymology. Therefore, to catch his thought was not easy for scholars without his background in Latin and Greek rhetoric (subject of which Vico was Professor at the Università di Napoli).

Vico conjectures in his *Scienza Nuova* that every civilisation exhibits in his evolution three recurring cycles (ricorsi), which he called «ages»: the divine, the heroic and the human.

He precisely states that each of these ages is characterised by distinct political and social features, which he exemplifies with specific «master tropes» or «figures of language». The «Giganti», i.e. Giants, are used during the divine age as a kind of metaphors to understand all natural phenomena. On the other hand, in the heroic age, the used tropes are metonymy and synecdoche, which model the development of feudal or monarchic institutions.

We recall here that the "metonymy" is a «figure of speech that consists of the use of the name of one object or concept for that of another to which it is related, or of which it is a part, as "scepter" for "sovereignty", or "the bottle" for "strong drink", or "count heads (or noses)" for "count people"» and that the "synecdoche" is «a figure of speech in which a part is used for the whole or the whole for a part, the special for the general or the general for the special, as in ten sail for ten ships or a Croesus for a rich man», as defined by Dictionary.com.

The feudal and monarchic institutions are leaded by some idealised superior figures, which are however perceived as human. The final age is characterised by popular democracy and the relative rhetoric trope is irony: in this final stage, the rise of rationality occurs.

However, Vico imagines the possibility of the occurrence of the «barbarie della reflessione», i.e. the barbarism of reflection, so that civilisations may descend again in their first or second stage. Vico talks about a «storia ideale eterna» or *ideal eternal history* constituted by the recurring cycle of described three ages. In his vision, this eternal history is «common to every Nation». Vico, being aware of the epistemological status of his theory, carefully describes the rise and fall of civilisations, using the conceptual tools which he has introduced and therefore provides evidence for his conceptual construction by the phenomenology obtained from the work of «descriptive» historians.

Many other observed facts may be interpreted with the use of the concept of Nash equilibrium.

Clearly, the position of Don Liborio was not lasting enough to call his agreement with Camorra as a stable Nash equilibrium: however the following history of Italy seems to indicate that such equilibrium has been attained in the systematic collaboration of the various kinds of Mafias with Italian (Southerner and Northerner, without distinction) politicians.

United Italy placed itself in a stable configuration where North regions economically dominates over South regions, by allowing to South criminality to dominate in some parts of the country: this is a Nash equilibrium which lasted more than 150 years. The main (and clever) feature of this equilibrium situation consists in the fact that there is no domination based on race or any other true or imagined personal feature of the involved individuals. On the other hand, there is nothing like a South Italian race as distinguished from a North Italian race, whatever is said by some supporter of some Italian parties and is written in the documents which US immigration office used in Ellis Island for selecting Italian immigrants. Once a South Italian has moved to North Italy or to the USA, he suddenly manages to become undistinguishable from the other citizens around him, except, maybe, when he shows to understand Neapolitan Language.

This freedom of movement and equality of personal rights is important for assuring the stability of the system introduced by Cavour to manage Italy: indeed there is a kind of tax imposed to those who live in South Italy.

If you want to live there, you must pay in terms of quality of life: worse health cares, diffused criminality and very bad systems of communication. Why then South Italy is not empty? Why many Germans and North Europeans want to live there? Because, it is really beautiful and climatically mild. Cavour introduced a kind of tax on beauty!

If you want to live in South Italy, you must be very rich, have many friends to support you and your family, as the State is often absent, and you must suffer, in any case, many hardships and deprivations.

For these reasons, it could happen that while Frederick II Hohenstaufen said: «I do not envy god for the paradise, as I am well satisfied to live in Sicily», Benedetto Croce wrote a book entitled: «A paradise inhabited by devils», instead (the paradise being South Italy!).

The social domination exerted in Italy is not based on a personal division between slaves and masters: it is, instead of regional nature. If one wants to go away from the exploited region, he can do it freely!

A dynamical equilibrium is then obtained: if the resources allowed to South Italians are not enough, many of them emigrate and therefore less individuals share the same amount of resources. If a rich man wants to live in South Italy, he goes there with his resources and makes the region richer. If, as happened during the period in which the Cassa del Mezzogiorno had some successes, the resources concentrated in South Italy increase, then many emigrants come back home.

Now, by using Nash Equilibrium Theorem, we can try to understand what happens to Southern dictators.

For some historical reasons to be better understood, the dictators in Southern regions polarised their features. Probably, the process is very similar to the process that leads to the development of peacock tails. Some Southern dictators are extremely precise, correct and respectful of laws. They are even too much respectful of laws and rules. Other Southern dictators instead simply ignore the existence of rules and are completely unscrupulous. They are too much unscrupulous. The more the dishonest dictators act illegally, the more the honest ones become stiffer in respecting the rules. The more the honest dictators, in their quest of perfection, become unsuccessful, the more the dishonest ones increase their criminal behaviour, pretending to believe that this is the only way in which some results can be obtained. The final configuration is the following: the many honest leaders of the South have the tendency to overestimate the strength of reason in controlling historical events and in driving history towards the best perspectives. Instead, the small amount of very criminal dictators take a great advantage in the naive behaviour of the many honest ones, and become richer and richer, more and more powerful, finally strangling their people.

The Southern dictators of the class of Don Liborio achieved ephemeral success indulging in crime. Southern dictators of the class of Carlo Filangieri achieved also ephemeral successes indulging in the idealistic belief that by pursuing the best they could ignore to achieve the good.

Northern dictators, instead, established among themselves another Nash equilibrium: they are all moderately bad. They follow and accept some basic rules and are all totally cynical in pursuing their own tactical and strategical interests. They, however, stop when their behaviour could too seriously damage their group interest. In this way, they were successful in using the weaknesses of dictators of the South (of both types).

What Northern dictators do not manage to understand is that with their accomplice behaviour they are allowing to Southerner criminality and bad social behaviour to become too powerful: they risk to have the whole Italian Nation be infected by the «asocial» structure which Cavour established in South Italy.

In a slogan: in Southern societies very few super criminal dictators parasitise too good dictators.

If this diagnosis is correct, one should wonder what will happen in Italy, or even in European Union, where the established free circulation of persons allows to Southern dictators to move North Europe and vice versa.

Will the system reach one of the two described equilibria? Will another intermediate equilibrium be attained? Practically: Calabria will move towards the Bavaria social standard or vice versa? Or an intermediate standard will be attained?

Mixed Random Strategies

Benedetto Croce in his «A paradise inhabited by devils» observed that very often some devils sometimes behave as angels. Many historians, while studying great men of the past, observed a kind of schizophrenic behaviour of many of them.

Frederick II Hohenstaufen polarised all chronicles into very positive and very negative ones; Carlo Filangieri and Cavour have been accused of great crimes and also cheered as great statesmen: the list could continue indefinitely. Even Don Pasquale has many admirers!

The question which arises now is: were these men definitely bad persons, and simply they managed to deceive somebody who believed that they were good fellows? Or they were good dictators whose deeds were distorted by evil-minded historians?

It seems, instead, that the conclusion is more complicated than that. All dictators were, at the same time, honest and dishonest, fair and unfair: randomly! Simply some of them were more often inspired by good intentions and others have mainly devilish behaviours.

To understand this statement, we must delve a little more in the details of Nash Theorem. Let us, therefore, specify a little better the content of Nash Theorem.

Let us go back to the concept of strategy, which we introduced before.

A pure strategy consists in a complete algorithmic and deterministic definition of how a player will play a game.

To be more precise, a player plays a pure strategy when the rule that she/he is following determines uniquely her/his move for any situation she/he could face in the game.

Following Nash, we define a **mixed strategy as an assignment of a probability to each pure strategy**. When a player chooses a mixed strategy she/he randomly selects, for every move, a pure strategy among the finite set of available strategies.

To make more precise our previous statement about Nash Theorem, we must say that:

Nash proved the existence of an equilibrium configuration for every finite game if the players are allowed to choose, in general, a mixed strategy.

Therefore, one can divide Nash equilibria into two subsets: pure strategy and mixed strategy Nash equilibria. In the first case, all players are playing pure strategies. In the second case, at least one player is playing a mixed strategy. Nash has also proven that not all games have pure strategy Nash equilibria.

It is now clear why so many dictators managed to give so different impressions to so many observers.

To get their power, they played the game by using a mixed Nash strategy.

In a percentage of moves, they were extremely fair, and in some others they were decisively unfair.

Here we want, as an example of a mixed strategy, to describe about the case of a famous scientist from South Italy, whom we will call Bianco Biondi.

Bianco in his life very often showed an extremely fair, honest and correct behaviour. He could be considered, at first sight, a follower of Carlo Filangieri strategy in the game of power. However, his career has been extremely quick, and his relationships with many powerful and most dishonest dictators of his academic group have been always extremely tight. When Bianco meets somebody, he spends a lot of time in listing the career occasions, which he has lost because of his honesty, always neglecting to describe the strategy that he had used to become full professor already at a very young age. Actually he has been for a percentage of 95% during his life an exaggeratedly upright scientist, always boasting his decency in social and academic behaviour. In reality, about 5% of his choices have been dictated by the determined and secretive support that he has given to one of the best allied of Don Pasquale. This allied used Bianco in few occasions to solve many blocked academic situations in which an apparently honest and fair judge was needed. Bianco, in change of his rare but very important services, obtained a full support by Don Pasquale and his allied for his own personal career, a support for getting academic visibility and the promise of an academic position for his children. Unfortunately, all of them did not want to pursue an academic career. In this situation, Bianco felt to have been deprived of a well-deserved reward and therefore looked for a kind of «personal attendant» who could help him in his everyday life, for instance typing his papers, writing his letters, buying what he needs for dinner in supermarkets and so on. Using his reputation as a correct scientist, Bianco managed to assure a professorship to his personal attendant: when a colleague asked him how it was possible that he had supported for such a prestigious position a person without any scientific originality the answer was astonishing: «I do not understand your morality, there are so many incompetents in our universities that one more will produce no harm. I need his services». In this way, at the end of his career, he has shown to somebody his true nature: it is for 95% the nature of an honest and clever scientist but for the 5% he is an extremely egocentric person, pursuing exclusively his own interests.

We think that the kind of phenomena discussed before and the mathematical problems to be solved for describing them should passionate the next generations of mathematicians, as these are exactly the most «useful» and «practical» problems which humankind will have to face soon. As Condorcet did during the French Revolution, also the mathematicians of the present generation will need to take their specific responsibilities towards future generations.

Is the Phenomenology of the Dictator and of the Leader Observed Only in Human Groups?

The main characters of this chapter will be Yeroen, Nikkie, Luit, Mama and Kanzi, chimpanzees' and bonobos' dictators whose behaviour has been described by Frans De Waal.

One could believe that what chimpanzees and bonobos are doing is not at all interesting for us. Instead, as De Waal is explaining so clearly in all his books, the observation of their behaviour is extremely useful to unmask the true nature of human behaviour and in particular to show the true nature of human dictators.

Indeed, chimpanzees' and bonobos' societies are a very useful simplified model for human societies, as the dynamics of dictator's behaviour are more transparent, when chimpanzees' and bonobos' troops are involved.

Chimpanzees do not try to find, as sometimes humans are doing, some a posteriori «logical» justifications for some peculiarities of their behaviour which, in reality, are innate. They are innate indeed, as these peculiarities are shared by humans with all the other apes and therefore cannot be a product of our superior intelligence.

A really important character of our narration will be **Kanzi**, the most famous bonobo, whose capacity of communicating with humans represents a challenge for social sciences.

No social and cultural superstructure is built to hide or justify the innate features of chimpanzee or bonobo ethology that we will describe in the following pages.

When one observes also in humans a behavioural feature described in apes, he probably is allowed to conclude that, for both apes and humans, it is innate and therefore their cultural interpretations, presented by humans, are purely justification excuses. Here, we will try to highlight as many similarities as we can, exactly to unmask a large quantity of human behaviours, which are often concealed with presumable very high and idealistic motivations, while being, instead, purely instinctive behavioural patterns.

© Springer Nature Singapore Pte Ltd. 2019
F. dell'Isola, *Big-(Wo)men, Tyrants, Chiefs, Dictators, Emperors and Presidents*,
https://doi.org/10.1007/978-981-13-9479-9_8

Mama, Testosterone and Dictators

Females of all species of apes must have an adequate level of oestrogen-type hormones to be able to give birth. Testosterone, the hormone that characterises males, is a hormone whose effects are antagonistic to those induced by oestrogen. Females also produce testosterone, but usually in small quantities: however testosterone is the primary cause of the libido of females.

In the group of chimpanzees described by de Waal and which inhabited the island of Arnhem Zoo, initially, there was no mature male. The most suitable individual to cover the role of the alpha individual (i.e. the group dictator, or in this case, the group big-woman) was, in that situation, Mama: a fairly strong and adult female chimpanzee. Indeed, as it was impossible for that group (as for every group of individual of social species) to live without anyone playing the role of dictator, Mama willingly agreed to take on the duty of regulating the conflicts among the other group members.

The biological effects of this social situation were immediate: testosterone production in Mama increased significantly, and Mama became somehow sterile. In fact, Mama had no children during the entire period of her group dominance.

De Waal observed, at that time, that Mama was not really able to effectively lead the group. Her reactions were "too violent", her punishments were "too harsh", and her actions were unnecessarily hyper-reactive and abrupt. At least in that circumstance and situation, a female was not really able to effectively assume the role of "dominant individual". Some feminists violently attacked the decision of the zoo's scientific responsible, when he decided to introduce into the group a male with the necessary skills to undermine the dominant position of Mama. Within a few months, and in a bloodless manner, thanks to the controlled intervention of the scientists, Yeroen took power. It has to be remarked that, Mama offered no little resistance to Yeroen's action, but, at the end, she withdrew into a role which was much more congenial to her: that of leader of the group of females, leader able to tilt the scale, in the struggle for power between males, for the one individual instead of another. She did manage systematically to choose as a leader of her group the individual who seemed most suitable to her.

Without the tension necessarily connected to the exercise of power, Mama completely forgot her previous violent behaviour, and indeed always preferred, among the males competing for the position of dictator, the least violent and most reasonable one.

Some time after Yeroen's arrival, Mama had the first daughter, but the testosterone level was probably still too high for her, and she was unable to take care of this child. She was "adopted" by another female chimpanzee, who was, by the way, also Mama's best friend. Only when Mama had her second daughter was she ready to take care of the infant, and the little daughter took advantage of her mother's high rank, to the point that the zoo scientists gave her the name: "Princess". As a high-ranking female, Mama's leadership had a very positive effect on the group: her actions stabilised it, her influence was essential to make decisions, her opinion was almost always the one which was followed by all the group members, including the dominant male. The murder of Luit by Yeroen and Nikkie, described shortly below, could only be perpetrated because in the night quarters of the males the females had no access, for a disputable choice of the group caretakers.

The phenomenology concerning the role of dominant females in social groups is observed both in human and in other primates groups and seems to be innate. Pinker (2008) accurately describes how females of our species "naturally" tend to manage power. Pinker believes that it is not true that our society refuses our women the access to the roles of power: her analysis leads to the conclusion that our women want for themselves a role of power and control over the lives of others which is "different" from what the males want for themselves. In short, in the opinion of Susan Pinker, adapting a female to a "male" place of power is well possible, but the female is unhappy, in the long run, of her life, and her way of managing this type of power is, in the end, inappropriate.

However, female dictators do like and want to exert power. In fact, they do not usually act directly, but through an intermediate individual.

This is what probably is doing the wife of Don Pasquale. In a sense, this is what Catherine the Great actually did. Catherine was, of course, the formal power holder: however she delegated the exercise of her power mainly to men, whom, for a reason or another, she believed were faithful to her person and/or to her political vision.

Yeroen, Nikkie, Luit: Male Competition for Power

Yeroen was introduced in the troop of chimpanzees in Arnhem Zoo for a specific reason. The experimenters needed a big-man for that social group, as the previously self-established big-woman had proven not to be suitable to manage power. To get the primacy in the pecking order, Yeroen needed to be helped by the experimenters: the alliances established by Mama with the other members of the group were too strong and even the greater physical force of Yeroen was not enough to change the situation. Mama was removed from the group for a while, so that Yeroen could start his persuasive action and could gain the needed alliances to withstand the confrontation with Mama when she was readmitted in the group. It has to be remarked that Mama did not renounce to his dominant position easily. She tried to keep her dominant role.

This is really meaningful in the context of the general study of power and its value for any individual: it seems that power is attractive for dictators independently of their gender.

The examples given by Catherine the Great and Mama are probably not anecdotic. Mama, in a sense, had experienced a unique life trajectory: she started being a formal dictator, directly exerting, without any kind of intermediation, power in her group. Subsequently, Mama tried to oppose to the establishment of the supremacy of Yeroen. At the end, she accepted to play the role of female's leader, conditioning every decision of the alpha individual: however she suffered when she was obliged not to be the first in the pecking order and to pay homage to Yeroen.

In her new social role, she acted as the king maker and the king controller of the alpha male in charge. This change of role allowed her to fully enjoy the happiness and responsibility of motherhood. This manifold experience made her a unique resource for her group, which recognised her authority and relied on her wisdom until the last minute of her life. The moving description of her last hours of life, surrounded by the affection of her human friends and her chimpanzee friends and relatives, reached the wide audience of the readers of some important international journals.

In the group, there was also another male: Luit.

The females of the group considered him handsome, they appreciated his temper and equilibrium, his psychological balance and his impartiality. He always and really enjoyed playing with children, while Yeroen clearly barely tolerated them, sometimes pretending to like to play with them, simply to please the females of the group.

As soon as Luit became confident enough to be able to challenge Yeroen, he started to dispute Yeroen's dominant role and to find the alliance of Mama and all the other females. For what concerns intelligence, capacity and suitability for being the group leader there was no doubt. Luit surpassed easily Yeroen, and rather easily he managed to become the alpha individual of the group. Yeroen had terrible depressive crises; he was always sad and susceptible and started to isolate himself from the social life of the group, not tolerating his rank decrease. For a while, the group could enjoy the enlightened leadership of Luit: this was the moment when Luit maximised the number of his descendants, as every female allowed him an easy sexual access. This moment of peace of the group and this sexual choice of the females assured to the birth of some of its most clever future alpha individuals: all children of Luit seems to have inherited his skills and revealed to be even more gifted than their father.

However, the evil, impersonated by Yeroen, was envious of Luit's successes.

Yeroen managed to imagine the possibilities opened to his power ambitions by the growing physical force of Nikkie, and by Nikkie's rather weak intelligence and discernment. Yeroen nurtured the ambition of leadership of Nikkie (also stupid individual have the innate instinct to become dominant individuals in their group!). Yeroen, by using Nikkie's physical force, managed to impose to the group a dual dictatorship. Mama, and all the females, did initially support the opposition of Luit to the new power regime. Unfortunately, Yeroen was too much determined, and when violence became a constant in the life of the group, Mama and the other females resigned themselves to the forced dual dictatorship. The females could not accept that their children be continuously harassed by Nikkie, with Yeroen who disturbed Luit so to prevent his active protection of the weakest. Nikkie was transforming, with the strategy of terror, the chimpanzee life into a nightmare: the females decided to give up and withdrew the support to Luit. Luit had, himself, a depressive mood and withdrew into a more solitary life. However, he never lost his dignity, differently from what had been done by Yeroen, who had cried, complained and exhibited tantrum crises after his power loss. The theatrical behaviour of Yeroen, showing a hypocrite facade, which was tried and exploited to touch the sensitivity of Mama and the other females, was never observed in Luit. Luit kept his emotive distance from the life of the group for all the time that was necessary to observe the establishment of a serious conflict between Yeroen and Nikkie.

Their agreement was clear to the experimenters: Nikkie was formally the boss, everybody in the group had to pay homage to him. However, Nikkie had to allow a complete sexual access to the willing females by Yeroen, and could contrast only Luit, and the lower ranking males, in this access. On the other hand, Yeroen was not intervening when a female alliance was protecting one female who did not want to accept Nikkie's advances, so that Nikkie's access to females was not unlimited. Finally, the formal leadership of Nikkie was really fictitious: the individual who was in charge of the conflict resolution and moderation was indeed Yeroen and only seldom Nikkie. This was a situation determined by Nikkie's incapacity to be really impartial: his friends were always preferred by Nikkie, in the decision concerning conflicts, with respect to those who were not his friends. This state of facts obliged Yeroen to intervene systematically to keep a reasonable level of harmony in the group.

When Nikkie decided that Yeroen could not access to a female, the crisis blew up. The equilibrium of power changed, also because Mama could not tolerate that the immature, violent and irresponsible Nikkie had an absolute power. The females decided that Luit had to come back to power and supported him decidedly.

This was the death sentence for Luit.

Indeed, he came back to power, literally elected by the females. Then, Yeroen and Nikkie had to experience the harshness of the power loss: they understood that sharing the power was more appealing than not having power at all. Finally, Yeroen conceived a putsch and put his plan into action. For a tradition established in every zoo, when chimpanzees are sent back to their night cages, males are separated from females and children. This separation was exactly what was needed to Yeroen in order to definitively get rid of Luit. Luit had gained an indefectible support from females: they had decided that Nikkie's brutal behaviour could not be tolerated. Therefore, Yeroen waited to be alone with Luit and Nikkie. Then, he kept Luit blocked to allow Nikkie to seriously injure their victim. Luit was found with some lacking fingers and some serious wounds. More meaningful and explicative, about the true intentions of the two criminal power seekers, is the main injury inflicted to Luit: he was evirated.

This fact needs an explanation: inside chimpanzee groups the physical confrontation is limited by a strong taboo.

Killing has been observed only very seldom inside a chimpanzee groups, while wars among different groups have been observed, leading the ethologists to conclude that the de-humanisation (better the de-chimpazeesation) process of the enemies is a mental mechanism which is shared between humans and chimpanzees. In other words, Yeroen and Nikkie had a strong taboo to overcome before killing Luit. However, they wanted to be sure that Luit could not come back to menace their power. The only viable behaviour that they could find was to remove his testicles: in this way, Luit push towards power and females was seriously impaired. Unfortunately, as a side effect, Luit died because of the injuries.

Unmasking Human Behaviour

Everybody in his life has experienced the abuse of power of dictators whose behaviours degenerate into tyranny.

All the scientists who interacted with Don Pasquale had the same feelings that were experienced by Mama, Luit, Nikkie and all the other members of the chimpanzee colony in Arnhem that has been dominated by Yeroen.

The study of chimpanzee politics and behaviour is extremely embarrassing for some humans, as it unmasks the true nature of human behaviour. Humans tend to hide their instinctive choices with elevated (and fictitious) motivations. Observing the same behaviour in primates definitively unmasks the truth. While we believe to be rational beings, instead, at least in the greater majority of our social interactions, we behave exactly as chimpanzee, or, when we show our better side, as bonobos (see the following pages about Kanzi!).

The arbitrary arrogant choices, the abuse of power, which Yeroen directly perpetrated or which, with his complicity, were allowed to Nikkie, had an emotive effect on the colony, which is very similar to those observed in human groups.

The frustration of Mama can be paralleled to the frustration of many serious scientists when they understood about the arrogant academic choices made by Don Pasquale. The most infamous ones are (i) his choice of an individual who cannot even be a good elementary school teacher as his successor as a department director in the secondary controlled University or (ii) his imposition of an illiterate person as a full professor in his prestigious home University, by manoeuvering the selection committees with the promise of promoting all its female members to a prestigious Academy. It is useless to say that only these unfortunate committee members had believed to Don Pasquale's promise, while everybody in the whole country knew that it was vain.

Many refuse to accept that chimpanzee behave exactly as Don Pasquale.

In facts in doing so, they should admit that, as Don Pasquale cannot, exactly as Yeroen, prove any theorem (also the simplest one) there is not a big difference between themselves and Nikkie, especially if they are allied of Don Pasquale. Also if the humans who are daring to oppose Don Pasquale, paralleling the conduct of Luit, may feel embarrassed, as they can see too many similarities between their own human behaviour and the one observed in chimpanzees' group.

Actually the following sentence

(…) an impulsive verdict: Yeroen was to blame.
He was, and still is, the one who decides everything in the chimpanzee colony.

which is an excerpt from the Chapter "Reflections on the dark side" in "Peacemaking among primates" by Frans de Waal (1989) can be adapted to the description of the academic events of the whole mathematical discipline dominated by Don Pasquale.

Replace in previous de Waal's sentence Yeroen with Don Pasquale and chimpanzee group with the Don Pasquale-dominated-scientific-group and it will reflect exactly also a situation observed in a group of humans, a situation which lasted until the final physical death of Don Pasquale and also beyond.

Indeed, Don Pasquale's power managed to prolong its effects several years after his death, as it is claimed that a selection committee did not dare to contravene his desires also then: they kept following his orders *usque ad mortem et ultra*, i.e. until his death and beyond.

Kanzi, the Bonobo: A Message of Hope

Kanzi's story begins with an overpowering act.

The dominant female in his group kidnapped Kanzi from his genetic mother immediately after birth. This episode in a group of bonobo in captivity is very reminiscent (see the pages about Catherine the Great) of what happened, at least once, in the Tsar's palace in St. Petersburg. While the most common crime in troops of chimpanzees and gorillas seems to be the killing (usually smashing their skulls) of infants, perpetrated by newly dominant males who want to be able to inseminate immediately the females whose motherhood burden has been violently removed, the kidnapping of infants seems to be the worse crime perpetrated by the dominating bonobos (i.e. the eldest females). It is clear, from the evolutionary point of view, the advantage obtained by newly dominant males because of the described violent behaviour, which, by the way, is observed also in lions and many other species.

It is not clear, on the contrary, which may be the evolutionary advantage obtained by dominant bonobo females in kidnapping other females children and in transmitting to them their high social status.

In humans, both behaviours have been observed. The infanticide of children perpetrated by the dominant male in a family is several times more likely if this male does not believe to be the father of the child. This could be one of the reasons for which DNA tests have shown that a percentage varying between 5 and 20% of human children are not genetically related to their social father: mothers, for protecting their children, hide the true paternity of their children. To this reason, one could also add the other reason that a man suitable to be protective and supportive usually does not have the dominant genetic phenotype, which is, instead, preferred by the primate mothers for their children. Therefore, both bonobo and human mothers try to use the strategy consisting of choosing the genotype of a specific man for their children and asking to another man, more suitable for this second purpose, to help in nurturing them. Hidden ovulation is presumed to have been developed by bonobo and human females to pursue this strategy (see, e.g. again De Waal).

The story of Catherine the Great, but many other similar stories, indicates that sometimes a dominant woman manages to become the owner of an infant's body and to replace its mother: this form of kidnapping is therefore present also in human ethology.

It is an open problem to understand why this behaviour is genetically advantageous, so that it keeps to be observed in species which are divided, in the tree of evolution, by a division occurred around three millions years ago.

Kanzi was born on 28 October 1980 and is a male bonobo whose behaviour has been described in many scientific studies concerning the intelligence and the use of language in non-human primates. Sue Savage-Rumbaugh, who is a famous and highly regarded primatologist and is supposed to be one of the most eminent scientist of her generation, claims that Kanzi exhibits, undoubtedly, advanced linguistic capacities. Kanzi was born to Lorel and Bosandjo, bonobos living in the Yerkes field station of Emory University and was moved to the Language Research Centre at Georgia State University few weeks after birth. He was stolen and nurtured by the dominant female of his bonobo group, whose name was Matata. Kanzi and his sister Panbinisha (who was genetically Matata's daughter) moved subsequently to the Ape Cognition and Conservation Initiative, in Des Moines, Iowa, where Kanzi has been immediately recognised to be the alpha individual of the resident community of Bonobos: this circumstance has to be remarked, as, in bonobo societies, it has been observed very seldom a male dominance.

Being nurtured by her, Kanzi was clung to Matata during the teaching sessions in which the experimenters tried to teach a symbolic language to Matata, through the use of a keyboard whose keys corresponded to lexigrams, i.e. symbols representing words. Matata did not show much interest in the lessons and the researchers started believing that their hope, i.e. to teach to a bonobo a formal communication code, was too ambitious and therefore bound to be proven as vain.

As a matter of chance, while Matata had been sent to another research centre to meet Kanzi genetic father to be inseminated again, the little Kanzi proved his outstanding skills and true temper. The absence of his mother caused him something close to a panic attack. Sue Savage-Rumbaugh was obliged to remain with him, all the time, night and day for at least three whole days.

Possibly also to prove to his new caretaker that he was worth being nurtured, Kanzi, the very day after the departure of his mother, proved to be able to use correctly 120 lexigrams. Subsequently, Kanzi learned several other hundred symbols, and their meaning include actions, objects, adjectives, names of familiar people and also some generic concepts (like Visitor, for indicating somebody whom he did not know before). Kanzi can also build two-words sentences, mastering also the different meanings related to the word order difference.

Kanzi needed to use words to survive, as Sue could not communicate with him easily in any other way: to get nuts, food, watermelon and any other thing which he needed, he used the code which he knew to be effective when dealing with human beings. Kanzi not only started using the lexigrams but also the chamber pot, which, when his mother was present, he had too often avoided to use. Kanzi's willingness to communicate with his caretakers and to follow the rules of education, which they demanded to him, has been interpreted, by the ethologists, as his way to show to his caretakers that he gave a great importance to their approval, which he perceived as essential for his survival.

His push to survival led him to become the first observed ape to have learned to use language by himself, without a training directed specifically to him.

This fact is even more surprising if one remarks that he has been also the first observed bonobo to be able to use a formal language at all. Remark that there is, also, evidence that bonobos can communicate among themselves, in a rather effective way. The interaction between Kanzi and Panbinisha has proven this fact: Kanzi did manage to help often his sister to answer to some of the experimenters' questions in a correct way, by using some vocalising typical of bonobos. Sue Savage-Rumbaugh reports detailed evidence of this fact: in one experiment, Kanzi was in a room separated from the room in which his sister could access to the lexigrams' keyboard. The vocal contact, instead, was not precluded. Kanzi was shown a yogurt and, after some vocalisations, managed to drive his sister, who could not see the yogurt, to the key meaning exactly «yogurt». This observation strongly suggests that those vocalisations, used by Kanzi and typically observed in bonobos, may have a precise meaning.

Experimenters working with Kanzi are aware of the potential criticism of sceptical scientists, who seem to be incapable to accept that a species sharing with human beings more than 98% of DNA can share with them, also!, some linguistic and logical capacities.

Therefore, to prove that Kanzi is really capable to recognise sounds and associate each of them to lexigrams, they speak a word in a microphone, while Kanzi is wearing headphones, in order to be sure to have filtered out all non-verbal clues. Invariably, when one of the words in his vocabulary is pronounced, Kanzi manages to point correctly the corresponding lexigram.

Bonobos societies are perfectly matriarchal: females alliances make the females of the group act in a strongly coordinated way and males are not trained by their mothers, and possibly not even genetically capable, to establish strong alliances with other males. The pecking order among females is automatically established via their age: seniority is the favoured criterion. In this way, social conflicts among females are avoided. The rank among males is a pure reflex of the corresponding rank of their (eventually adoptive) mothers. However, when the mother dies then his son's rank collapses and a highly ranked male, simply because he has lost his mother, becomes the omega individual in the group. Matata has been, during her life, the alpha individual of her group. There is therefore no surprise to remark that Kanzi's rank was correspondingly very high.

Surprisingly enough Kanzi kept his high status also after the death of his mother.

This is a serious challenge for ethology and mathematical modelling of primates' behaviour. Kanzi has kept his dominant posture and conduct, signalling his social status, during all his life, at least up to now, as he is still alive. He has serious, expressive and calm eyes, he is balding and has a paunch. He resembles many human big-men, Chiefs, Presidents and Faculty Deans: his behaviour recalls the interlocutor, in every circumstance, that he is reliable, wise and authoritative.

Now we can dare to make some conjectures, which will need to be confirmed on the basis of a deeper and more careful analysis of phenomenology and the development of predictive theoretical models.

First of all: possibly the evolutionary advantage of Matata genotype in kidnapping a clever boy is not irrelevant.

Panbinisha, and most likely all the descendants of Matata, can profit the talents of Kanzi. The time and resources investments spent on Kanzi by Matata are not an evolutionary waste: instead they acquire to the Matata's lineage a very convenient support. The evolutionary advantage so obtained may be relevant and may justify the kidnapping and adoption. One has to remark how different are the dominant female strategies and the male dominant strategies. Dominant males destroy genetic and blood resources smashing skulls, a behaviour which is very inconvenient for the species as a whole. Instead, dominant females steal the best «genetic» resources to adopt them in their own lineage. These are both forms of violence, but while the first endangers the survival of the whole species by favouring a specific lineage, the second one while favouring a lineage is not harmful for the whole species.

Secondly, we must observe that in bonobos the dominance of males is extremely rare. In a sense, such dominance is unnatural.

Exactly as unnatural, in chimpanzees, is the dominance of females. However, the bonobos group where Kanzi is living managed to recognise his intellectual superiority, and therefore chose him as an alpha individual: indeed he is clearly, the most capable to bargain with human beings the best possible treatment for the colony.

This was a very wise and democratic decision, which goes against any habit, tradition and customs.

Possibly, it goes also against genetic tendencies, written in bonobos' DNA. However, it is convenient, very convenient, and therefore it was adopted. This is possibly the major manifestation of the intelligence of bonobos.

This, for sure, is a message of hope, also for humankind, as the peaceful behaviour of bonobos is most likely written in our DNA, also. Sure, this peaceful behaviour is in contrast with a similarly natural Chimpanzee and Gorilla male dominant, violent and conflictual inheritance, which we also share with the other two our closest species (*see De Waal, Our inner ape*).

However «Nature, is what we were put on earth to rise above.» as, wisely says Hepburn's Rose Sayer in a famous film. Moreover, resorting to the bonobos' part of our nature, this rise may be easier.

Before concluding, we must confute some objections of some sceptical scientists that blame all the researches on Kanzi.

They claim that the observed facts are anecdotal and not reproducible, and therefore scientifically meaningless. We must admit that Matata was not capable (and in great part his genetic daughter Panbinisha, also) to have the same intellectual performances as Kanzi. Therefore, the aforementioned sceptical scientists claim that Sue Savage-Rumbaugh reports have no scientific meaning and that, in particular, they do not prove anything about the intellectual capacities of primates.

There are many of these kinds of anecdotes, we list here some, which were, for us, more surprising.

Kanzi participated, when he was eight years old, to a research program aimed to measure his capacity to understand spoken requests and to make a comparison with this ability as shown by a two-years-old human child named *Alia*. Both Kanzi and Alia were given 660 spoken instructions. These instructions were asking the two children to deal with familiar objects, but in a novel way. Kanzi performances were slightly better than those of the child: 74% of successes compared with 65%. Remark that giving the exact number gives to the story a much more scientific and objective appearance!

We think that this «anecdote» is teaching us a lot, independently of the numbers. It shows us how to deal with our children. It tells us how limited are, at the beginning, our children's capacities, but also how big development potential they have. Having learnt to cultivate the capacities of Kanzi is teaching us how to deal with our own children to get the best out of them.

Kanzi managed to learn to build by himself and to effectively use some stone cutting tools. These tools have been compared to Oldowan tools. Kanzi needed a training period, his teacher was an expert anthropologist, but eventually his stone knives proved to be very sharp and effective. Kanzi could develop his own method for flaking, different from the one he was taught to use: he is reported to have produced 294 artefacts.

During a picnic, Kanzi indicated the symbols for «marshmallows» and «fire». As reported by Susan Savage-Rumbaugh: «Given matches and marshmallows, Kanzi snapped twigs for a fire, lit them with the matches and toasted the marshmallows on a stick».

Kanzi manages to mimic some sound of human speech: he is doing it notwithstanding the big physiological differences in the vocal apparatus between humans and bonobos.

A researcher, required to do so by Savage-Rumbaugh, performed a haka in presence of the whole troop of bonobos led by Kanzi. As it is well known, haka is a Maori war dance, whose meaning is fundamentally aggressive: it is the Maori version of a chimpanzee, gorilla or bonobo display. As such, it was perceived by all bonobos, except Kanzi. While his companions responded in a similar aggressive way, Kanzi remained perfectly calm and communicated somehow, using his vocalisations, to Savage-Rumbaugh his intentions. He wanted to see the dance again, but not in the presence of other bonobos. The performance was repeated in a private room and Kanzi could have his curiosity satisfied.

Kanzi can play the Arcade game Pac-Man and manages to beat it. As Kanzi enjoys very much eating omelettes, he manages to cook them by himself. Obviously, he uses his lexigram keyboard to ask for the ingredients.

We believe that these anecdotes have a precise meaning and interpretation: Kanzi has the intellectual capacities of a human child.

Linguistically, he is skilled maybe like a child of three years old, but his leadership proves that other aspects of his intelligence are comparable with adult humans. The sceptical scientist asking for reproducibility may be answered in the following way, which is paraphrasing De Waal:

> If an alien intelligence arrives to the earth and, by matter of chance, meets Einstein, it can conclude that humankind can master general relativity and possibly nuclear energy. However, as also in alien intelligence the sceptical philosophy has been developed, it will be afraid to conclude that his statement is true. Indeed by asking to several billions human beings it will discover that Einstein is an anecdote. Therefore he will conclude that theoretical physics is outside the human capacities.

The message of hope transmitted to us by the simple existence of Kanzi is simple.

In human genes, at least those that we have in common with bonobos, there is what is needed to select a clever leader, the most suitable one for his group.

In human genes, there is also a great attitude towards the peaceful solution of social conflicts.

Phenomenology of Servitude

Soyez résolus à ne plus servir et vous voilà libres.
Be determined not to serve anymore and you will be
instantaneously free.
Étienne de La Boétie in The Discourse of Voluntary Servitude.

In this chapter, we want to discuss a masterpiece of political thought: the famous pamphlet by Étienne de La Boétie entitled *Discours sur la servitude volontaire*. In order to get an English text to use for our citations, we have referred to the translation by Harry Kurz, which we have found particularly faithful of the original.

In the edition of La Boétie work commented and introduced by Murray N. Rothbard, one finds some titles for each of the three parts in which it is divided. We will comment the ideas of La Boétie correspondingly and using the understanding which we have gained by means of Condorcet conjecture, Arrow's theorem and modern evolutionary ethology observations and results.

It is surprising how careful, lucid and analytic is the analysis presented by La Boétie, even when he leaves some questions partially answered or totally unanswered. It is lucky that Montaigne, his close friend, did manage to transmit to us La Boétie's complete manuscript.

The titles of the parts that we reproduce below are a very synthetic and meaningful *résumé* of La Boétie's arguments.

Part I. Why Do People Obey a Government

The fundamental political question is why do people obey a government. The answer is that they tend to enslave themselves, to let themselves be governed by tyrants. Freedom from servitude comes not from violent action, but from the refusal to serve. Tyrants fall when the people withdraw their support.

© Springer Nature Singapore Pte Ltd. 2019
F. dell'Isola, *Big-(Wo)men, Tyrants, Chiefs, Dictators, Emperors and Presidents*,
https://doi.org/10.1007/978-981-13-9479-9_9

The question raised by La Boétie is the most fundamental question if one wants to scientifically design a democracy.

It is clear that there exists a tendency of a human being to follow the orders of the set of individuals in power, set which La Boétie calls collectively: «the government». If one wants to design a democratic institutional system, he has to understand at first the social, psychological and cultural mechanisms which, acting together, induce people, in general, to obey.

«Why do people obey …?» The answer which La Boétie gives can seem, at first sight, nearly a tautology: «they tend to enslave themselves».

Actually we can try to enlarge this argument, with the understanding we gained with the considerations presented in this work, as follows:

(1) Every social group needs continuously to decide which action is needed for its survival.
(2) Unfortunately, the debate among the individuals of the group cannot produce easily (recall the Condorcet conjecture) a social choice.
(3) Therefore, only those groups, in which a dictator, in a way or another, has been chosen, did manage to increase their chances of survival.
(4) In a long evolutionary process, only groups that developed an effective system to select leaders did survive and in nature only this kind of group is presently observed.

Remark that we observe leaders in groups of very «simple» species, for instance, in groups of hens. Therefore, the structure of societies based on leadership has been selected by natural selection very early in the history of life, on earth.

Probably, humans tend to enslave themselves, exactly for the same reason for which humans tend to close their eyelid when they believe that somebody is striking them: it is an ancestral instinct.

A schizophrenic attitude is, however, present in humans, apes and many other species: every individual who sees the smallest possibility to get power tries to reach it.

Indeed, another ancestral instinct appears widely diffused: the tendency to try to overpower the competitors to the position of dictator of a group whatsoever. It is not important how unimportant is the leadership role that can be contended: any leadership position always finds more than one pretender. In the same individual, a slavery instinct coexists with a huge ambition towards power, independently of the eventual presence of the awareness of one's own limits.

Also when a self-pretending dictator knows that he is not suitable for the office to which he strives, he finds a justification to his ambitions. Usually the argument reduces to only one: there are so many incompetent leaders that one more will not be of true harm.

The slavery itself is justified in the same way: if everybody bends his head, why should I risk facing powerful people by rebelling? In this way, slavery is justified by the impossibility to change the subjection situation.

Freedom from servitude comes not from violent action, but from the refusal to serve

Gandhi or Mandela personified this prescription by La Boétie. It is true, in general, that the most effective way to free oneself from servitude is to place oneself in a position in which it is possible to refuse to serve, even if this may have some costs. Apes are, as a group, much weaker than human. As individuals, however each ape, may it be gorilla, chimpanzee or even bonobo, is extremely stronger than a human. In general, more than five strong men are needed to overpower physically a chimpanzee. On the other hand, apes are very well aware of their inferiority with respect to humans, their fear of and respect for human is proven without any doubts. However, apes innate refusal of servitude did not allow humans to use them as indefatigable workers. Apes are genetically incapable to be slaves.

Apes prefer to be all killed instead of bending to be domesticated. Humans, instead, are clever enough to be able to accept servitude to be able to survive. Therefore, the first followers of la Boétie must be the ancestors whom we have in common with other primates!

Tyrants fall when the people withdraw their support.

History of humankind is full of examples of tyrants whose delirium of omnipotence led them to their ruin.

Yeroen was obliged to surrender his power to Luit because of his greedy behaviour both the first and the second time. The subjects (i.e. the group of females) withdrew their support.

Don Pasquale knows this fact so well that he never ignores the demands of his subjects: probably, because of his wife, he never acted in such a way that the group of incompetent scientists who accepted to be his subjects would even conceive to withdraw their support.

Caesar Sacristy resembles much more Yeroen than Don Pasquale: he has lost his power several times, because his being clearly unfit to become a well-liked dictator. As Yeroen, Caesar, is extremely gifted in succeeding in the fight to get power. Instead, both of them are unfit to be leaders. Therefore, they lost and regained power in different moments and different forms, keeping straining towards power until their physical death.

Part II. Liberty Is the Natural Condition of the People

Liberty is the natural condition of the people. Servitude, however, is fostered when people are raised in subjection. People are trained to adore rulers. While freedom is forgotten by many, there are always some who will never submit.

In this part of La Boétie more precisely describes the innate schizophrenic nature of humans (and apes, as the reader will discover by perusing De Waal textbooks).

Aspiration towards liberty is nurtured by the evolutionary advantage obtained in becoming dictators.

Acceptance of servitude is nurtured by the evolutionary advantage obtained in accepting other individual dictatorship, once a personal advantage has been bargained.

We disagree with the literal acceptance of La Boétie belief that people «are trained to adore rulers». Instead, they are descendants of innumerable individuals who survived simply because they could accept servitude. Only the extreme overconfidence of humans on their own intelligence and on their capacity to dominate instincts may lead somebody believe that it is possible to get such a systematic submission, as that observed in human societies, simply by training.

Submission is clearly an acquired phenotype of many species, including apes and humans. Freedom is not forgotten! We simply claim that it is convenient for many to live in servitude, and that therefore natural selection wrote in our DNA, our capacity to accept servitude.

However, it is true that there are some humans who will never accept to submit: these individuals are the seeds of competition towards power in humankind. Their existence has been left among us in order to assure to human groups the capacity to renew their leaders, when the leaders in charge become dramatically unfit to lead their people.

The reader should think to American and French Revolutions, when the desire of freedom became uncontrollable, simply because the dominant classes did not manage to interpret the changed aspirations of their subjects. It was exactly during French revolutions that the ideas of La Boétie were rediscovered once more. Marat used them as the starting point of his analysis while Condorcet did start the analysis that we hope to have clarified a little bit in this book.

Slavery is the preferred status of incompetents who cannot accept their incompetence.

There are many ways in which incompetence manages to exploit its slavery: some are rather unexpected. There are cases in which incompetent persons manage to induce competent scientist to become slaves. Don Pasquale could acquire the help of Bianco Biondi, who is a clever scientist (however, it has to be said that he has a tendency of appropriating the ideas of others without recognising their contribution).

How Don Pasquale managed to do so? Simply by exploiting Biondi's need to have a personal attendant.

Unfortunately, this attendant had no mathematical or scientific competences. He is totally incompetent, although he has the ambition to become a professor. Moreover, he does not want to admit to be incompetent even to himself. Therefore, after having tried to produce some scientific results, in vain, he decided to impose to Bianco his ultimatum: either you find a professorial position for me or I will not serve you anymore. To be served for some more time, Bianco became a servant: the servant of Don Pasquale and his wife. His ethical justification was the usual one: there are so many incompetents that one more will not be of harm.

Incompetents hate competent people and fight against them.

Incompetents perceive competence as an insult. Indeed immediately after having got his tenure, the personal attendant did not even answered to the phone calls of Bianco. Another, not expressed, justification of Bianco for his own unfair behaviour was therefore proven be to vain. Indeed, Bianco believed to be such a genius that he deserved a personal attendant paid by the taxpayers. However, the incompetence, after having exploited the «ego» of the competent, usually rebels.

We conclude our considerations with some wise words:

[...] dove men si sa, più si sospetta. (v. 66)
[...] where [or when] the less is known, the more it is suspected.
Dunque, non sendo Ingratitudin morta
ciascun fuggir le Corti e' stati debbe;
che non c'è via che guidi l'uom più corta
a pianger quel che volle, poi che l'ebbe. (vv. 184–187)
Therefore, not being the "ingratitude" dead,
everybody must escape the Courts of the States;
as there is not a shorter way, which guides the man
to lament that which he wanted, than having obtained it.
Excerpts From Dell'Ingratitudine (On Ingratitude) by Niccolò Machiavelli

Part III. The Foundation of Tyranny

If things are to change, one must realise the extent to which the foundation of tyranny lies in the vast networks of corrupted people with an interest in maintaining tyranny.

Caesar Sacristy and Don Pasquale are motivated in their search of power by envy and ingratitude. However, the most relevant reason for which Caesar Sacristy and Don Pasquale did manage to keep their power is to be found in the general attitude of their subjects. These slaves found the realisation of their aspiration to freedom via the attainment of their miserable personal profit and their little apparent advantage via the subtraction of resources to the community. Moreover in the search of their miserable profit, they also enjoy the disgrace and the impediment inflicted to the action of those whom they know are more capable. Therefore, we can talk about an alliance of the envious, which needs the choice of a leadership: those who excel in envying and in being ungrateful.

Both Caesar and Don Pasquale have chosen to select the members of their academic group with the criterion of incompetence, lack of morality and stupidity.

In this way, they found themselves at the centre of a network of immoral interest whose satisfaction was possible only via the diffusion of corruption. The dominant and majoritarian group of individuals unfit to their assumed commitment towards highest culture and scientific specialisation has, therefore, an indefectible interest in maintaining the tyranny.

La Boétie describes in a vivid way the courtiers, who

> have to think what the King wants them to think, and often enough they must anticipate his thoughts in order to please him. It is not enough for them to obey, they have to please him. Serving him destroys them, yet they are expected to share his joy, to abandon their tastes for his, to change their nature and constitution. They have to be attentive to every single one of his words, to the tone of his voice, to his gestures and facial expressions. Their eyes, feet, and hands – everything has to be ready to read the mind of the King and to satisfy his wishes.

Even if absolute monarchy disappeared in Western nations, in the voluntary courts of Don Pasquale, Caesar, and many other small dictators, the behaviour described by La Boétie can be still be observed.

The main feature of the voluntary slaves consists in their apparent and declared honesty. They always criticise those who want to oppose to tyrants, denying that the methods used are the most appropriate. They do not participate to the fight against tyranny because «they are honest» and «other are the methods to be used»! Often, they spend time to justify what is unjustifiable. Indeed,

> *La vertu s'avilit à se justifier.* (Virtue is debased by self-justification.)
> Voltaire Oedipe, act II, scene IV (1718)

The described fake and specious honesty resembles very closely the brittle behaviour in materials, as described by mechanical sciences. Brittle materials are very stiff under very small loads, but when greater loads are applied then their sudden fragile rupture occurs.

The voluntary servitude is based similarly on the brittle morality of many humans: they apparently show to be very honest, strict and respectful of laws. Under small external moral loads, they are very stiff: they reproach often to other persons a lack of morality, when they blame small, and often completely unimportant, infringements of rules. Instead as soon as their interests are involved, then they undergo a sudden morality rupture. In this situation, they are ready to sell their mother, their children, their best friends, their lovers, their brides and grooms or their benefactors for getting the smallest personal advantage.

Like all the dictators of their species, Caesar Sacristy and Don Pasquale are basing their own power by the systematic exploitation of this kind of «brittle» honesty.

Instead, resistance to tyrants must be based on resilience and ductility. Stiff behaviour must be replaced by adaptive ductility. In material science, this property consists in a material's ability to undergo significant plastic deformation before rupture. Therefore, a ductile material manages to sustain greater loads without undergoing collapse. In social situations, anybody is subject to great loads: the capacity to change his own behaviour, attitude and opinion, by keeping steady principles, ideals and actions, is essential. Of course this implies adaptivity: exactly that adaptivity which is criticised by the tyrants' voluntary slaves, who pretend an indefectible honesty, being instead ready, beyond a certain threshold, to accept any kind of immoral compromise. Often these compromises are also useless, as the tyrants are very dexterous in getting favours without paying them back. However, they do this in a very selective way: they repay back some favours, to keep safe their credibility in these deplorable exchanges, but they capitalise a great part of the obtained favours without repaying them back, to get a net personal advantage.

As suggested by La Boétie, the only way to stop the power of the dictators belonging to the class so beautifully exemplified by Don Pasquale and Caesar Sacristy is to stop having any interaction with them, by refusing any contact and compromise. Unfortunately, this strategy is not easy to implement, as its success requires the support of a large percentage of group members.

Actually, it seems that there is a phenotypical structure of apes and humans behaviour that leads some individual to «instinctively» support these kinds of dictators.

It has been suggested that some kinds of bipolar depressive paranoid disorder and also some kinds of schizoid personalities have been developed in human and apes societies in order to give the individuals showing these phenotypes an evolutionary advantage.

Servitude as an Innate Instinct

We try to argue more precisely the reason for which we believe that so many personality disorders were developed. They seem to be a behavioural adaptation needed to compensate the distress induced in many individuals by the unavoidable establishment of the pecking order in human social groups.

The question, which we try to address now, is

Which could be the biological reason why many humans are prone to develop a respondent conditioning that oblige them to «obey» to malevolent dictators?

We believe that the vision of Jakob von Uexküll (a famous biologist who lived between 1864 and 1944) about animal ethology can be extended also to describe some «apparently elevated» aspects of the human behaviour. The starting point of von Uexküll is the concept described by the German word «Umwelt» which can be translated into English as «surrounding world». Uexküll claims that each living being considers reality exclusively from his point of view. Consider for instance the eyeless ticks. These living beings are clutched to a grass stem even up to fifteen years in a kind of suspended life, in the total absence of nutrition. They wait to perceive the chemical signal produced by the butyric acid emanating from mammalian skin. If it will not eventually die, the tick manages to drop onto a mammalian and ends its period of fast by filling itself with warm blood. Then, it lays its eggs and, being its life cycle ended, it dies. Of course the tick does not mind, as it is not capable to conceive this consequence, that inside its body other parasites may be waiting to infect the mammalian, which even is supplying the so vital warm blood. The mammalian not only gives a drop of its blood to a parasite, but also is eventually infected by another parasite, with the complicity of the tick, which, nevertheless, has been beneficed.

Which is the perceived world (Umwelt) of a tick?

It seems very limited. However, its limited perception of reality is also the essence of its terrible strength. A tick has a well-defined objective and nothing can distract it. For the tick, the Umwelt consists into three and only three experiences:

(i)   The perception of butyric acid,
(ii) The perception of the temperature of 37 °C (i.e. the temperature of the blood of every mammal),
(iii) The perception of the presence of hairs of mammals.

Nothing else has a meaning, nothing else is perceived, nothing else exists for a tick.

Similarly, a person suffering for a schizoid personality disorder cannot even conceive the concept of affection, intimacy or emotional involvement.

For them, these feelings do not exist and, in general, schizoid personalities doubt, in general, about the possibility that somebody could really feel these sentiments. Indeed, they do not show any interest in social relationships, have a solitary or sheltered lifestyle, never share their thoughts with anybody, are secretive, emotionally cold, detached and very often apathetic. Schizoid personalities cannot have friends, allied or any form of intimate attachments to others: many schizoids manage to simulate these feeling while pursuing their own «vainglory» dreams. Indeed they have an elaborate and exclusively internal fantasy world, in which, usually, they imagine to have a dominating role in a purely fictive network of unequal relationships. The intellectual Umwelt of schizoid personalities consists of a series of imagined performances, which they try to transform into reality in order to satisfy their internal distress.

In general, the concept of Umwelt refers to the genetically determined self-centred, subjective world of an organism: it is the part of the world that matters for the survival of the species. It is a very small part of the «reality», but the enormous remaining part must be ignored, as perceiving it is not producing any evolutionary advantage. Uexküll proposed to map the Umwelt of all species to give a key for understanding the nature of living being. Some philosophers claimed that understanding the Umwelt of other species is beyond human possibilities of comprehension. Others claim that a member of a human society cannot understand even the Umwelt of other social groups. Instead, we believe that the biggest advances of human science have been achieved when our understanding managed to attain the point of view of other living beings.

Only recently scientists managed to understand the Umwelt of persons suffering of mental disorders.

In particular, the study of mental disorders greatly advanced when it was possible to enter in the inner world of personalities who, for instance, suffer for a **narcissistic personality disorder**.

These personalities have an exaggerated feeling of self-importance, an outmoded demand of admiration and a total lack of empathy with other human beings. To understand that a narcissistic personality simply cannot understand that, in reality, he is not the most admirable person in the world is essential, if one wants to try to care this disorder.

Another really painful mental disorder is bipolar disorder. It is characterised by periods of depression and periods of self-inflated too elevated mood. In the elevated mood, the individuals rarely sleep and suffer of episodes of as mania or hypomania. During mania (and in hypomania, with reduced and milder intensity) episodes, an abnormally energetic, happy or irritable behaviour is observed, and the concerned individuals systematically do not ponder carefully their decisions, by simply ignoring the consequences of their actions. Mania usually presents manifestations of increased self-esteem or grandiosity, rapid speech, the subjective feeling of rapid and clever thoughts, disinhibited social behaviour, senseless impulsivity and episodes of persistent paranoid behaviours and attitudes. Instead during the periods of depression, one can observe senseless crying, an unjustified negative perception of life, poor eye contact with others, persistent (and not-related-to-reality) feelings of sadness, irritability or anger, loss of interest in previously enjoyed activities, excessive or inappropriate guilt, hopelessness, sleeping too much or not enough, changes in appetite and/or weight, an unjustified feeling of fatigue, problems in concentrating, feelings of worthlessness. In the most severe cases, one can observe serious psychotic symptoms like delusions and hallucinations.

Also, to look for a relief during the depressive episodes, a **self-serving bias** is manifested in the individuals' behaviour. In this psychotic behaviour, all the cognitive and interpretative actions are distorted in order to maintain and enhance self-inflated esteem. It is associated to the attitude of perceiving oneself exclusively in a favourable way: the suffering individuals believe that their successes are exclusively related to their own capacities and engagement and are sure that their eventual failures are exclusively due to external actions of often malevolent persons or fate. These individuals refuse negative feedback, repeat continuously to themselves «as good they are», as high is their scientific standing, how great is their research, their strengths and their achievements. They systematically refuse to consider their responsibility in their failures, overlooking their faults. When they admit faults, they believe that their failures are due to the lack of quality of their team's work. However, while sharing the common work with other team members they tend to avoid the tasks, which they are incapable to fulfil. The reason for which they behave in this way is that they need to protect their self-inflated ego from threat and injury: when perpetuating illusions and error, they satisfy their insuppressible need for self-esteem.

Actually bipolar personalities show a high risk of suicide and self-inflicted harm: it has been estimated that there is a strong genetic basis to the onset of this disorder in more than the 80% of the observed cases. Indeed, in some recent brain imaging investigations, it was possible to detect some great differences in the volume and structure of some specific brain regions in patients exhibiting a bipolar disorder when compared with the brains of healthy control subjects.

Don Pasquale and Caesar Sacristy are genetically capable to exploit all these disorders.

They manage to manipulate their court of voluntary servants, by giving ephemeral satisfaction to their psychoses and compensative actions.

The most successful courtiers of Don Pasquale and Caesar are the schizoid personalities. Their lack of emotions is very useful: a schizoid is a perfect courtier. For him, insults and humiliations simply do not have any apparent effect. The schizoids obey until they get what they wanted. Then they simply abandon the court. A schizoid waits like a tick even for years: when the conditions are favourable to the satisfaction of their inner impulse, they suddenly change their behaviour, often completely ignoring those previous personal relationships that they established simply for reaching their self-assigned aims.

Both the two academic dictators we have studied in detail (i.e. Don Pasquale and Caesar) have a very complex and manifold personality. It is sure, however, that their narcissistic vision of life is one of their dominant temper features. Maybe, this is the reason why they can manipulate so easily both narcissistic voluntary slaves and those personalities characterised by a dominant self-serving bias. Especially, Don Pasquale is capable to satisfy with very miserable recognitions the greatest narcissistic egos, often by using them as a weapon against his enemies. Voluntary servitude is too often accepted in change of the venting of narcissistic neuroses, as the dictator who is being served well will protect the narcissistic persons from the obvious revenges usually implied by the egotistical and arrogant behaviour that they need to act for being satisfied.

However, where the most successful dictators really excel is in exploiting bipolar personalities. Actually, we could say that a kind of symbiosis is established between narcissistic dictators and bipolar voluntary slaves. These dictators have a rare manipulative capacity: when the bipolar personalities are *down,* they are not able to oppose to a strong order of their dictator. Actually, they accept the orders because: *I am a peaceful honest person and I cannot start stupid power battles.* When a bipolar personality is *up,* then their grandiose attitude can be easily directed towards the opponents of the dictators: actually it is easier to attack somebody who is in a weaker position. In this way, bipolar personalities manage to have a neurotic satisfaction of their distress being, also, helped by the powerful dictator in charge. They even manage to develop a sincere gratitude towards the dictators for having allowed them to vent their mental rage by participating to their dictatorial miserable power plots.

The question is why the described personality disorders did not disappear because of natural selection?

They seem to be a rather negative phenotype for the survival of human kind!

The answer to this question is very similar to the one that can be given to explain why thalassaemia is not disappearing in human groups where malaria is an endemic disease.

Indeed mental disorders can be regarded as strategies, taken to extremes, both for becoming dictators and for surviving to the action of dictators.

To make the situation more complex, one has to observe once more that, unfortunately and in general, the talents needed to an individual to become a dictator are not the talents, which a dictator needs to lead his group towards survival and success.

Natural selection keeps choosing successful individuals, as they are favoured in the sexual competition and eventually manage to have more descendants: therefore the existence of mental disorders must have some evolutionary positive effects!

The case of Genghis Khan is exemplary, in this aspect. His paranoid behaviour, his violent treatment of subjugated peoples together with his careful attention to the well being of his own people are the evidence of how complex must be the temper of a successful dictator. As a matter of facts, to be prudent, and to be capable to understand the good or bad intentions of other individuals, is a very positive talent for a dictator: more, it is very beneficial not only for the ambitions of the individual who wants to become a dictator but also for the group lead by such a dictator. However, if prudence, and awareness of the evil which is hidden in human nature, becomes paranoia, then the dictator becomes dangerous for his group and most likely he will lead it to the self-destruction.

Taking a positive phenotype to extremes is never a successful survival strategy.

The example evoked few lines before, and taken from medicine, is very useful to understand this point. We talk about thalassaemia. It is a terrible disease, which kills, with terrible torments, children after few years of life. The obvious question arises: how it is possible that this disease is so diffused in Mediterranean Europe, Africa and Asia? Why this phenotype is always present in the human population if the children suffering of thalassaemia usually do not survive enough to reach the reproductive age? The corresponding genotype should disappear immediately after its occurrence in some individuals. The explanation of this phenomenon has been understood only recently. Related to thalassaemia, there is a very beneficial genotype whose corresponding phenotype, i.e. microcytemy, gives the capacity to resist to malaria. More precisely, microcytemy is a feature of red blood cells (it is a peculiar shape which does not allow to the carrier of malaria to infect the organism), which is very advantageous. However, if both mother and father are carrying the gene of microcytemy then, following Mendel rules of transmission of genes, there are 25% of chances that their child, inheriting a pair of genes of microcytemy, may suffer of thalassaemia, i.e. a faulty capacity of haemoglobin synthesis.

Thalassaemia is a form of microcytemy taken to extremes.

The evolutive advantage gained by populations in which microcytemy genes is diffused is very large if they must survive to endemic malaria. The fact that, in these populations, a percentage of children is seriously ill can be seen as a side effect which is less deadly than the action of malaria agent in the populations without the microcytemy gene. This gene, when not doubled in the DNA of a child, corresponds to a beneficial phenotype. When, instead, it is doubled it corresponds to a deadly phenotype.

A kind of Nash equilibrium is attained in the interested populations, with families using a mixed strategy. Not all the individuals carry the gene of microcytemy, but only the percentage, which minimises the aggregated amounts of deaths because of both malaria and thalassaemia. Too much microcytemy genes mean too many deaths for thalassaemia, not enough microcytemy genes means too many deaths for malaria!

The attentive reader will have already understood why we are giving this example here: too much distrust on other human beings actions and intentions leads to paranoia, i.e. the false belief that in every situation other human beings are behaving in a hidden non-cooperative or aggressive way. As a consequence, a dictator falling into paranoia may destroy his group or lead it to a disastrous situation. On the other hand, a dictator believing too much into the good intentions of human beings will either be dethroned by his competitors inside his group or will cause his group ruin, by allowing to the group's enemies to subjugate or destroy it.

In the competition to transmit one's genes, human beings (and primates) are trying many strategies. These strategies require different skills. These skills are some kinds of phenotypes, generally corresponding to a specific set of genotypes. This correspondence, the role played by nurture and the importance of nature in establishing this correspondence, is not fully described and understood yet. Therefore, we will limit ourselves to intuitively try to describe the positive effects, which the discussed mental disorders may carry, if not taken to extremes, to the humans suffering because of them.

A bipolar disorder may produce a very positive output in an academic career as governed by Don Pasquale.

If the candidate suffers for a years-long-depression episode, he is considered a docile instrument by Don Pasquale, who cannot perceive the dangers of his manic episodes, which usually occur only when the depressed individual has already got a permanent position. Or, like a tick, the bipolar individual may spend years in a frozen psychological state, waiting when somebody will give him a possibility to have a career, possibly by destroying using delation or any other malevolent and immoral behaviour the career of other persons, usually exactly those who helped the depressed individuals when they were suffering the most.

Of course, bipolar depressed individuals very often manage to have success in their career or in their struggle to survival: if their depression is not too severe, they manage to transmit their genotype to future bipolar depressed individuals.

Moreover, Don Pasquale is very efficient in exploiting the manic episodes of long-term depressed individuals, exciting their self-esteem. These individuals during manic episodes show a paranoid attitude towards those who had helped them the most: they look for a justification for their betrayal. The Umwelt of bipolar personalities is completely distorted: they believe to be the only truly honest and fair persons in the world, while they behave in a completely cynical and opportunistic way. They cannot say what exists outside their internal world and they wait, like the tick, for the smell of the blood of the persons whom they need to parasitise. A bipolar voluntary servant manage to betray those who had helped them the most, in the hope to get support from Don Pasquale.

A schizoid personality has even a bigger evolutionary advantage: not having any capacity of affection they are capable of every kind of action to pursue their psychotic objectives and are powerful instruments in the hands of Caesar Sacristy and Don Pasquale. They simulate affection and make people believe in their reliability: however when their interest calls them to the worse abjections, they can forget everything and «go ahead». Actually Don Pasquale managed to use the wife against the husband, one brotherly friend against the other, the pupil against his Maestro. When somebody is schizoid, he can do everything, as he has no sentimental feedback to control his actions.

Of course if a schizoid personality behaves too much, and too openly, cynically then its evolutionary success is seriously impaired.

What has to be understood is that a bipolar personality, a schizoid personality, Don Pasquale, Caesar Sacristy, the servant of Biondi, they all have their own Umwelt and that they really do not understand, perhaps even they do not have the conceptual tools for perceiving, a reality which is outside the range of their limited «senses».

This last staement is true albeit some human beings lineages do have developed so-called «mirror» neurons.

They are also present in non-human apes and, in a simpler form, also in lower primates. An interesting discussion on these neurons and their role in the formation of abstract thought can be found in (Acharya and Shukla 2012). The system of mirror neurons provides the «hardware», i.e. the physiological basis, for the mechanisms of perception/action coupling. Strong similarities are found, in this biological context, with many concepts of the theory of codes: mirror neurons have been developed for allowing humans to understand the action of other individuals and for learning by imitation. Mirror systems seem to allow our brains to simulate parts of reality and for potentiating our language abilities. Mirror neurons allowed human being for the increase of our capacity to have a precise representation of the surrounding world and for enlarging our «Umwelt». Unfortunately, this increased capacity becomes a terrible weapon in the hands of Schizoid personalities: like ticks, they are not capable to really mirror in their brain the feelings of others, but they do manage to simulate empathy. After having established apparently strong personal links, they exploit the empathy of those who believed to be loved (or supported or understood) in order to get the satisfaction of their psychotic needs.

Which are the disorders at the basis of the life-lasting behaviour of the puppets masters, Don Pasquale and Caesar Sacristy, it is too difficult to say: they are too diabolically clever to show the full extent of their twisted brain.

However, the considerations expressed by Machiavelli in the epigraphs about ingratitude quoted before, and a paroxysmal expression of psychotic envy, jealousy and inferiority complexes together with a narcissistic ego may theoretically explain the reasons for which they can exist, notwithstanding the big harm that they cause to our species.

Of course, disturbed personalities showing too severe disorders are cut out by the natural selection. Only the less severe or mild versions of these personalities survive, to offer to dishonest dictators their voluntary servitude.

Montesquieu: The Possible Democracy

For concluding this essay we want to explicitly state that the search for democracy should not be considered a hopeless endeavour, as it is proven by the evolution from the populistic Athenian Democracy into the Roman Republic and then into the modern Constitutional Democracies.

Towards a Rational Theory of Constitutions

There are possible solutions to the Condorcet/Arrow impossibility results!

As we know very well from everyday life, and how many mathematicians have showed in human history, there are problems that have no solution. In particular, the problem of finding a function of choice, which reflects the demands of democracy, as formulated by Arrow, has no solution.

Can we conclude from Arrow's theorem that democracy does not exist?

In a very long period of human history, from the dictatorship of Julius Caesar to the French Revolution, it was believed that democracy could not work. Even without the Arrow theorem, Western philosophers have always theorised the inevitability of monarchy. The great majority of them believed that the system of rules, which European monarchies had given to themselves, was sufficient to prevent them from turning into despotism.

When the founding fathers of the American Constitution formed their utopia of democracy, all European intellectuals believed that the North American system could not work, since until then only aristocratic-monarchist systems had managed to survive and prosper. However, both the system established by the Constitution of the USA and the system developed, in its various forms, by the French Republic did prove to be very effective. How was this possible?

© Springer Nature Singapore Pte Ltd. 2019
F. dell'Isola, *Big-(Wo)men, Tyrants, Chiefs, Dictators, Emperors and Presidents*,
https://doi.org/10.1007/978-981-13-9479-9_10

In mathematics, very often, the impossibility theorems have opened new horizons of research and have produced very important and new theoretical developments.

One of the most ancient and classic example of this circumstance is given by the impossibility theorem formulated by the Pythagorean school: there is no rational number, that is, there is no fraction, whose square is equal to 2. The Pythagoreans believed that all numbers were fractions: this result led to the conclusion that no number could measure the hypotenuse of an isosceles right triangle whose catheti measure 1. Clearly, this paradoxical conclusion cannot be accepted. How was this paradox solved? Extending the concept of number. A real number is a succession of fractions that approximates better and better the quantity that it characterises. In short: the problem of the determination of the length of the said hypotenuse is solved with a procedure based on successive approximations, and by inventing a new mathematical concept.

The set of real numbers, that is the mathematical model of the physical concept «quantity», is not constituted only by fractions but it also includes sequences of fractions. These sequences, by means of successive approximations, more and more precise, come closer and closer to the quantity considered.

Remembering the solution of the Pythagorean paradox concerning the root of 2 can help us to face also the problem to which we are turning our attention now. That is to say:

How can we solve the problem of designing a democracy, given that a single «democratic» function of choice does not exist?

We could say that this problem can probably (and hopefully) be solved as Charles-Louis de Secondat, Baron of La Brède and Montesquieu had imagined: by means of a system of successive approximations!

The Division of Powers Imagined by Montesquieu

The Spirit of Laws (*L'esprit des lois*), published in 1748, is one of the major theoretical contributions to political thought in the history of mankind. It has been defined as the encyclopedia of political and legal knowledge of the Age of Enlightenment. Obviously, it was harshly criticised by Jesuits and was placed on the Index (*Index Librorum Prohibitorum*) in 1751 also because of a negative judgment given by the professors of the Sorbonne.

In book XI of this essay, the theory of the separation of powers is formulated for the first time in modern philosophy of law.

Since "**absolute power corrupts absolutely**", Montesquieu poses the problem of analysing the nature of power, in order to distribute it, in its different forms, to different independent «dictators».

Montesquieu distinguishes, at least, three kinds of the different «powers» that must be exerted in a State: the **legislative power** (which consists in stating the laws), the **executive power** (which must work to solicit and enforce the laws and manage the *res publica*) and the **judicial power** (which must judge those who violate the laws, to block their future actions). Moreover, **Montesquieu states that a necessary condition for the exercise of the freedoms of every citizen is that these powers be clearly separated and independent of each other**.

Montesquieu did not believe that a certain constitution or state organisation can be valid for every society and at every time. However, as Giambattista Vico does, he is convinced that, despite the diversity of particular events, history manifests via a constant order and immutable natural laws. Although the rules necessary for the proper functioning of a State are not absolute, that is, independent of space and time, and must be changed as situations change, Montesquieu believes that there are fundamental principles to which a society cannot derogate in giving itself laws, if it does not want to go to ruin.

These principles must provide for the distribution of the ultimate power of choice in various kinds of decision-making process to different independent agents. These specific kinds of decision-making processes must be carefully determined and must include: formulating and enacting laws, taking the necessary action for the well-being of a society and preventing criminals from their illegal actions.

We conjecture that all the despicable behaviours described in the previous phenomenology of the dictator can be prevented or, at least, greatly decreased in quantity and quality if the groups there described were subject to rules inspired by the Montesquieu principle of separation of powers. Obviously to be applicable, this principle requires that every kind of power be explicitly attributed to specific decision-making bodies via a precise formal procedure.

To interpret Montesquieu ideas in the context of the analysis consequent to Arrow's theorem it is necessary to find different sets of social choices and it is necessary to find different functions of choice, for each of these sets.

For instance: a function of social choice will be needed for establishing the laws, another function will be needed to guide the government action and another one to rule the action of judicial power. For each of these functions of social choices, it is necessary to identify, in a precise way, a dictator (or a college of magistrates acting as a dictator by majority votes) who must be able to decide, if necessary, in the name and on behalf of the State. However, several different dictators must be vested with such ultimate decision-making power depending of the subject concerning the decision. Moreover, all these dictators must be independent of one another and their role must be clearly and precisely specified *a priori*. The number of independent dictators must be sufficiently large, the scope of their decision-making power must be well defined and their role must be limited in time and space. Finally, each of them must be guaranteed total independence of judgment and freedom of choice.

To define exactly what we mean with the expression «democratic system» is of fundamental importance if we want to try to build it using mathematical constructions.

In mathematics, you need to know how to ask the right questions if you want to find meaningful answers. The art of asking the right questions is difficult. This art consists mainly in the choice of the right mathematical definitions.

In Arrow's theorem, we try to characterise a democratic system by using the concept of the function of choice and imposing for it some conditions, which assure its democratic nature. Unfortunately, this brilliant idea has not led to a solution of the «practical» problem of basing on solid mathematical principles the writing of a constitution and the laws framed therein, being thus sure of their efficiency, applicability and truly democratic nature. However, Arrow's theorem has made us understand the reasons why, inevitably in every human or animal society, the figure of a dictator has always appeared. Understanding the fundamental reason for this experimental observation not only has given us a rational comprehension of many phenomena but also has disclosed many new problems.

One of these problems, perhaps the most important one, is:

How to organise the complex constitutional division of powers in a society in order to be sure that this society is capable to choose the best for itself as a whole?

As Vico and Montesquieu have already clearly argued, it is the set of laws established and respected that make societies healthy and one cannot hope that individuals invested with decision-making power will use it in the interest of the whole society (if the laws allow them to pursue their own exclusive interest).

Therefore, the previous problem can be reformulated in this way:

Which constitution and which laws must be formulated to ensure the most effective organisation of a human society? What are the fundamental principles to which constitutions and laws must conform in order for the societies they govern to be successful?

More Sophisticated Mathematical Theories Are Needed

Mathematicians who are interested in social sciences must dare to design a society by means of a more sophisticated mathematical theory than that developed by Arrow.

Instead of using the «simple» approach based on Condorcet's concept of function of social choice, we must try to formalise, mathematically, the approach based on Montesquieu's thought.

Here we try to give an outline of the mathematical theory that we would like to develop.

We start with the set of «complex» choices that must be made by the decision-making system, which oversees the life of the considered society. A complex choice **sc** is an ordered n-tuple of choices (**s1**, ..., **sn**). We call SC the set of all possible complex choices. Therefore, the social utility is a function U which associates to each SC element the real number U(**sc**), which represents the utility for the society obtained with the complex choice **sc**.

One has then to solve the following problem: **How to determine the mechanism that leads to adopt the «best» complex choice sc?** Obviously, we want the choice adopted to be as close as possible to the choices that maximise social utility.

It should be noted that each individual has his own function of utility: let us call it Ui. This function is also defined in SC. Let scM be a choice that maximises U and let scMi be the choices that maximise each of the individual utility function Ui.

Anyone who has studied a little of the history of humanity has observed, without any doubt, that scM usually does not coincide with any of the scMi.

Therefore, a single dictator in general is not inclined to pursue the common utility, that is, the scM choice. The reader may think, for example, of Genghis Khan's choices, as described in the chapter dedicated to him. Obviously, his ultimate goal was to have the widest possible genetic progeny. He only dealt with the welfare of his subjects in an instrumental manner, in order to pursue his main objective. It is probably for this reason that his empire has had an ephemeral duration, while his DNA, and in particular, his Y chromosome, is widespread among humans today. Genghis Khan provides a simple example of a dictator whose function of utility does not coincide with that of his people.

Similar considerations can be done when thinking of the examples given by Don Pasquale and Caesar Sacristy. When conceiving, using mathematical concepts and models, a constitutional system one must always remember how specific dictators behaved, in order to check the efficiency of proposed institutional systems with the particular examples that are known.

The set of social choices SC must then be partitioned with a suitable relationship of equivalence: each social choice must belong to a certain subset and this belonging must be precisely determinable with a clear algorithm. The SC set must therefore be regarded as the finite union of SCj subsets (with j varying in the set of the first M natural numbers: 1, 2, ... M), M being the number of different decision making context, each requiring a different independent «dictator». The idea that can be taken from Montesquieu's speculations is as follows: for each SCj, an independent decision-maker must be designated.

Such a dictator can be an individual or a collegial body or a diarchy formed by an individual and a collegial body or two collegial bodies.

We give few examples here.

In the Constitution of the USA, it is provided that the Supreme Court makes the last decision on the constitutionality of laws. In this case, the dictator is a collegial body. The members of the Supreme Court are appointed for life, to ensure their total independence from the other powers. They are appointed by a diarchy: the President proposes the appointment to the Senate, and the Senate must approve it by a qualified majority.

In many jurisdictions, the figure of the monocratic judge is provided for: in this case, the dictator is a single individual, but his decision can, in general, be appealed. In bicameral legislative systems, a law, to be effective, must be approved separately by two chambers. The method of establishing the Italian Constitutional Court is subject to a larger set of checks and balances. The constitutional judges last only nine years in office, and one-third of its members are elected by the Parliament, with a qualified majority, another third of its members are higher magistrates elected by electoral colleges formed by magistrates who hold certain offices and the last third are chosen by the President of the Republic. In any case, Italian constitutional judges must have the highest academic and legal qualification.

In all the examples given, the Constitution provides for a precise determination of the roles of the various dictators involved in the final decision to be taken. Each dictator can only decide on well-defined choices and the Constitution provides all possible guarantees to safeguard their independence.

Therefore, one can try to formalise the idea of Montesquieu as follows: to find the partition of SC and the method of designation of Dj dictators for each decision-making field SCj such as to maximise U, taking into account that each dictator will maximise Uj.

To our knowledge such a formalisation and the corresponding theorem has not yet been presented in the literature, but we are confident that, by developing, among some others, the methods of the Nash equilibrium theorem, it is possible to obtain a result of this kind.

Appendix A
The Story of Two Words: Dictator/Tyrant

Words, and their meaning, evolve over time.

Since the Second World War, the word dictator has taken on a negative meaning, which it did not originally have. Let us briefly see which kind of evolution it had. Indeed, the history of the word dictator shows important aspects of the phenomenology of social groups aspiring to democratic government. Over the centuries, these societies have experimented with various forms of organisation and structures, in order to be able to compete with other societies and to last over time.

The invention of the role of the dictator in a republican constitution has to be attributed to the Roman constitutionalists, the word dictator having been invented in Latin.

Surely, Roman Republic has been one of the most successful attempts of Contitutional Systems in terms of duration and quality of life of its citizens. For this reason, the institutional solutions used by the Romans deserve special attention and must be studied with the spirit used in mathematical physics.

The word dictator has not always had a negative meaning, negative meaning which, instead, has been always given to the word tyrant.

The word tyrant has been first introduced in ancient Greece and in modern dictionaries has the following meaning: «a ruler who seized power without legal right»: and this meaning has surely a very negative content.

To make the situation a little bit more complicated, one has to recall that the history of Syracuse (Greek colony in Sicily) has been positively influenced by some of his tyrants: Hieron II, most likely, was a wise ruler, who kept under his protection Archimedes.

Dictator comes from the Latin «Dictator» derived from the verb «Dictare».
It is a reinforcing verb of «dicere» (i.e. «to say»), which means «to command». The verb *dicere* is also the root from which the word *dictum* was formed, which translates into command, order (think also of the English verb *dictate*). The dictator

© Springer Nature Singapore Pte Ltd. 2019

F. dell'Isola, *Big-(Wo)men, Tyrants, Chiefs, Dictators, Emperors and Presidents,*
https://doi.org/10.1007/978-981-13-9479-9

represented the "sovereign magistrate" and must be considered a characteristic figure (perhaps the most characteristic) of the constitutional order of the ancient Roman Republic.

The dictator was elected by a specific procedure (from the consuls in office together with the Senate) and for a fixed period of time.

His election was required at times of greatest danger to the Republic. The dictator had unlimited power both in war and in peace: according to some sources, he was called so because what he dictated, that is, what he ordered, was law. No magistrate of the Roman Republic had equal powers, but the very sources that attributed to him the «summum» imperium, that is the «supreme» imperium, implicitly indicated the limits of this power when they specified that the dictator cumulated in his person the prerogatives and duties of both consuls. The Constitution of the Roman Republic, like the modern British one, was not formalised in a precise document and can be considered a first example of a flexible constitution. It is therefore difficult to establish whether the dictator of the Republican era was a magistrate, an extraordinary magistrate or whether he assumed a «monarchical» role similar to the Spartan kings. It is, however, certain that Roman law accepted the twofold need to identify specifically and with great precision the cases in which it was allowed to appoint a dictator and to limit in a precise manner his prerogatives and powers.

In Roman Republic, a dictator was appointed for a specific reason and had a specific task to accomplish.

It would seem that the Constitution of the Republic, in ancient Rome, while taking note of the fact that in many situations in democracy the recourse to a dictator is inevitable, sought to regulate the appointment and functions, especially by limiting the duration of his special powers in a strict manner.

The dictatorship was then discredited by what can be called the «coup» of Silla. Indeed after Silla, only Caesar accepted to be elected dictator, and the consequences of his choice are known! Before Silla, the only specific tasks to which a dictator could be called were: to quell a revolt, to govern the State in difficult situations (**rei gerendae causa**), to convene the comitia (i.e. the gathering of electoral bodies) for the elections, to plant the clavus annalis, the annual nail, in the wall of the temple of Jupiter, action necessary for the calendar calculation of the years, to determine the festivities, to officiate the public games, to celebrate certain processes, to nominate new senators in the Senate.

The Roman legal tradition has inspired many legislators, statesmen, and politicians who, in later periods, have had to deal with emergency situations.

It will be noted that Roman law has tried to circumscribe as precisely as possible the role and tasks of the dictator: an admirable attempt to prevent abuses and *coups d'état*.

The inspirers of Roman constitution, while realising the enormous advantage that the Republic derived from the possibility of naming a dictator in critical moments, were perfectly aware of the risks that endanger democracy because of this institution. Actually, since in the flexible Roman constitution it was not possible to formally provide for a strict and effective control of the dictatorial power, the Roman Republic institutional functioning was substantially disrupted because of the consequences of the actions, in sequence, of two dictators: the conservative Silla and the progressive Caesar.

The most famous dictators who did save the Republic in times of difficulty were Cincinnatus (who save Rome during the war against Aequi, in his first dictatorship, and stopped a *coup d'état* during his second dictatorship) and Quintus Fabius Maximus Verrucosus, surnamed Cunctator, i.e. the "the delayer" (during the second Punic war against Hannibal).

During the civil war between *optimates* (some powerful noble families and their oligarchic supporters) and *populares* (other powerful noble families and their plebeian supporters), Silla marched on Rome and was elected by the electoral comitia «dictator in order to constitute the Republic and to write laws» (dictator rei publicae constituendae causa et legibus scribundis). This new dictatorship did not correspond to the traditional one, since it had no time limit and the election did not emanate from the Senate and the two consuls. Silla held this office for years before voluntarily abdicating and retiring from public life.

The analogy between the political conduct of Silla and that of De Gaulle, held many years later, is very suggestive.

In the presence of what De Gaulle called a "parliamentary dictatorship" (i.e. an institutional structure in which the power of veto of the parliamentary minorities, in an extremely divided Parliament, paralyses any possibility of action, condemns governments to instability and generates a chaotic policy), De Gaulle assumes the role of Prime Minister of the Fourth Republic with the specific task of promoting the change of the Constitution (he becomes Prime Minister *rei publicae constituendae causa et legibus scribundis*). With decisive actions (and probably trespassing the limit of the republican constitutional legality), he succeeded in completely reforming the institutions, arriving at the formulation of a new Constitution: what has been called the Constitution of the Fifth Republic. He used (illegally, when considering the Forth Republic Constitution which was in force) in an effective and repeated way the institute of the popular referendum (paralleling Silla's, recourse to the vote of the comitia). Having realised that his reformer role had been exhausted, he resigned from his office as President of the Republic before his term had expired.

Julius Caesar modified the dictatorship *rei gerendae causa* bringing its duration to a full year.

Appointed for the year 46 B.C. as dictator for the management of the Republic, he could be elected nine times in succession to this office, which was initially intended as annual, so becoming dictator for ten consecutive years. Finally in 36 B.C., the Senate voted to appoint him as «dictator perpetuus» (perpetual dictator). It was Caesar himself who, by modifying the institute of dictatorship, made the word take on, for the first time, a negative meaning. After Caesar's assassination, the consul Mark Antony had the dictatorship abolished and expunged it from the republican constitution. Later, the office was also offered to Augustus, who prudently preferred to combine in his hands several elective magistracies. Augustus was thus a dictator (in the new sense given to the term by Caesar) only «de facto», without formally admitting it. Augustus, instead, preferred to be called Emperor.

Dictator and the derived word dictatorship are learned terms: taken from the legal language of Latinity, they entered the English language with the first translations of Latin texts. The words dictator and dictatorship have then gradually taken on a negative meaning, increasingly linked to that of the word tyranny.

As Machiavelli writes, "the first tyrant to be in that city, he commanded it under the title of dictator". Machiavelli uses the word dictator in a positive way, but suggests that there is a strong tendency for a dictator to become a tyrant. As we have already reported, the word tyrant, of Greek origin, was born with an intrinsically negative meaning: this negative meaning remained always attached to the word.

Therefore, the careful use of the words dictator and tyrant should impose a necessary distinction between concepts that it is useful to keep distinct: a dictator is not a tyrant.

The word tyrant comes from the Latin tyrannus, which in turn comes from the Greek τύραννος : as we already said, it means "he who holds the power illegitimately". And as we have said, Caesar, Silla and, more recently, De Gaulle have reached the position of dictator in a legitimate (or apparently legitimate) manner. Silla and De Galle, too, were able to give up their power when their reforming role was over. We do not know what Caesar would have done, as he was killed by some of the many Roman citizens who were considering him a tyrant.

With an unstoppable historical drift, the words "Dictatorship, dictatorship, dictator" have assumed over time an almost exclusively negative value.

This negative meaning is not limited to political contexts: in the eighteenth century, for example, dictatorial was used in general with the meaning of despotic, authoritarian. With the spread of democratic ideas in the period of modern revolutions, dictatorship came back to indicate a form of government characterised by the concentration of power in one person, without any intrinsic negative meaning.

Indeed, the figure of the dictator continued to have many positive functions: for example, during the main phase of Italian unification, between 1859 and 1860. In that period, in fact, some patriots, and in particular Garibaldi in the Kingdom of the Two Sicilies, took the title of dictator in the name of King Vittorio Emanuele II and while awaiting its annexation to the new Kingdom of Italy.

Unfortunately, the dictatorial power, which was assumed in order to deal with situations of real danger to democracy and to resolve real and serious problems, was often exercised in a tyrannical and anti-democratic manner. This seems to be intrinsic in human nature.

Many dictators got their power not only by using their control of the military apparatus, but also often by obtaining the support of the popular masses. To this aim, they skilfully used demagogic propaganda and the establishment of a single party to which all citizens, which aspire to get public offices, must belong.

The twentieth century has seen many dictatorships of this kind, such as the Fascist, Stalinist and especially Hitler's dictatorship, which have caused enormous suffering to the peoples of Europe. It should be noted that both Mussolini and Hitler came to power in a completely legal manner. Only after having reached the top of the political structure of the democracies, only then did they operate a series of *coups d'état* and finally became tyrants. Exactly as had been forecast by Machiavelli.

The just-concluded discussion proves that it is appropriate to «nickname» Arrow's theorem as the theorem of the dictator.

A dictator exists only if there is a system of rules operating to perform social choices. His role is conceived to overcome the deadlocks in the process of forming a social choice. Unfortunately, a dictator can be tempted to use his absolute powers to attain his own personal interest instead of that of the society that nominated him. Arrow's theorem states that as deadlocks in a social choice function are unavoidable, a dictator is, unfortunately, always needed for breaking them.

The reader should not conclude, however, that democracy is impossible.

As Demagogic Athenian democracy was improved by Roman Republic and then evolved into modern constitutional democracies, the concept of a unique function of social choice has to be overcome. Our rudimental ideas in this context were discussed in the chapter where the ideas of Montesquieu are presented.

Appendix B
More Details about the Mathematical Structure of the Theorem of Dictator

We believe that some more details must be given about the logical structure of the theorem proven by Arrow.

However, we decided to present them in a final appendix. The reader, after having seen the enormous potentialities of Arrow's analysis, should be motivated to confront some conceptual difficulties. However, we are not addressing the present pages to a specialist. Indeed, here we expect that the reader knows only the basic concepts of high school mathematics and we will try to explain them again, giving as many explanations as possible. The reader is simply assumed to be a curious intellectual: somebody who knows some parts of knowledge very well, who is interested in abstract thought also when it is applied in other disciplines and who wants to understand the intrinsic logics of natural phenomena. Obviously, we also hope that some young reader will be attracted by the predictive power and the intrinsic elegance of mathematics and may be prompt to study more deeply it.

B.1. Individual or Social Choices Among Several Options

Let us start by considering a social group G made up of N individuals. Let **i** denote the generic of such individuals. In other words, we use the letter **i** to indicate one element whatsoever of G. Mathematicians say that **i** is a variable. The possible values of the variable, which we consider here, can be a generic element in the set of integers between 1 and N. We are assuming that every individual of G has a name, and we have chosen a natural number between 1 and N to be one of the possible names.

Let us suppose that the group G must express its preferences, as a whole, by choosing from a set of conceivable alternatives or options (as a synonym of the word «alternative» we will also use the word «options»).

We will call S the set of all conceivable alternatives. Each element of S is an alternative which is viable for the group G in a given decision-making process.

© Springer Nature Singapore Pte Ltd. 2019
F. dell'Isola, *Big-(Wo)men, Tyrants, Chiefs, Dictators, Emperors and Presidents*,
https://doi.org/10.1007/978-981-13-9479-9

Let us give an example: G could denote the set of all the American citizens during the Cuban crisis. One could imagine that the set of alternatives S which were viable consisted exactly in the following options, which all together form, in the present instance, the set S:

(1) Pretending not to know that the Soviets are sending atomic missiles to Cuba,
(2) Threatening economic sanctions but not initiating military action,
(3) Block access to Cuba with its own fleet to prevent the arrival of Soviet weapons and let the Soviets know that the blockade would be maintained even at the cost of a naval battle between American and Soviet naval units,
(4) Attack Cuba immediately,
(5) Attacking directly the Soviet Union,
(6) Sending more nuclear missiles to Turkey in retaliation.

Obviously, each individual in G has his own personal preference among the options listed in S.

The problems we want to study here are the following: (i) how to represent mathematically the choices made by each American citizen among the six options? (ii) how can we process the set of all choices made by all American citizen in order to obtain as a result the choice of the USA as a whole social group?

To make precise the analysis of the problem of Condorcet, we must introduce a mathematical concept that is an effective model both of the choice of each individual in G and of the social group G, taken as a whole. In order to do this, we must first of all introduce the concept of a total order relation into the set S of all the possible alternatives.

Definition (*Relations in S*) Let us consider a set S, and let us denote with the symbol (s, t) any pair of elements of S. **A relation in S is established** when we can univocally determine the list of pairs of elements (s, t) for which the relation holds.

Example The relationship «to be preferable for citizen **i**». In the set of options listed before, a certain citizen **i** may decide that: (6) is preferable to (5), (1) is preferable to (2), (3) is preferable to (4). The pairs $((6), (5))$, $((1), (2))$ and $((3), (4))$ are pairs which are in relation, in the considered case.

The reader will understand easily that a shortcut in notation is needed, if we want to avoid lengthy and confusing sentences.

If instead of writing (1) is preferable to (2) we write $(1) > (2)$, the meaning of the written text is the same, but the number of involved symbols is smaller.

Actually, we can say that mathematical formalism consists simply in finding the best possible symbolic representation of used concepts. The introduction of the notation $(1) > (2)$ is an example of the mental operation performed by mathematicians in analysing the logical structure of the theories.

One could also decide that (4) is equivalent to (5). If one wants to include this case, the relation to be introduced must include the possibility that one option may be preferred or may be equivalent to another option. The definition, which will be formulated below, extends in this aspect the previous one.

Ordering a set of options is essential for expressing a choice. A choice must specify, for each pair of options, which is the preferred one. Arrow claims that a **choice** is obtained in a set of options when for every pair of options, one can say if they are equivalent or if one of them is preferred in the given choice. The following definition makes this concept precise.

Definition We will call «total order relation in S», a relation, denoted with the symbol $\leq$, between elements of S that verify the following properties.

(**Totality**) For each pair (x, y) of elements of S, one and only one of the two possibilities occurs: 1.1 $x \leq y$ (i.e., x is related to y) or 1.2 $y \leq x$ (i.e., y is related to x).

(**Antisymmetry**) For each pair (x, y) of elements of S if $x \leq y$ and $y \leq x$, then $x = y$ (i.e., x is equivalent to y).

(**Transitivity**) For each triad (x, y, z) of elements of S if $x \leq y$ and if $y \leq z$, then $x \leq z$.

This definition may seem very abstract: somebody could even say that it is «too» abstract. In reality, it gives a precise description of a class of mental operations to which we are all very well accustomed. Of course not all choices, in the meaning given to this word in the standard language, are choices in the sense of Arrow. It is possible that an individual cannot decide himself about his preference between two options. For instance, it is possible that a citizen may not be sure about his preferences when dealing with options (1) and (2) in the Cuba crisis scheme given before.

Therefore, mathematicians add to a word which is used in standard language (in this case choice) another word specifying better its meaning (in this case, the added word is an attribute: Arrow).

If an **Arrow choice** in a set of options is made, then for every pair options (x, y) the involved decision-maker can decide either that x is less preferable than (or equivalent to) y (in symbols: $x \leq y$) or that y is less preferable than (or equivalent to) x (in symbols $y \leq z$). Following what Arrow wants to mean with the word «choice», in every choice a decision-maker MUST decide which one of the two options in any pair is preferable for him.

Remark that: **if it happens that $x \leq y$ and $y \leq x$, then the two options are equivalent for the decision-maker (in formulas x = y).**

When using the symbol $x \leq y$, we will say indifferently that x is equal or smaller than y, that y is preferred or equivalent to x, that x is related to y, that x is equivalent to or less preferable than y, when the given relation is considered.

If one has a set of possible options S, then his choice, in the sense of Arrow, must be a total order relation in S.

Let us ponder again on this point: what does choice really means for Arrow? Once a choice is made, then it has to be said that, for example, option 4 is preferable or equivalent to option 5. Not only that: between any two possible options, it is necessary to say which is the preferred one or if they are equivalent (**Totality of the**

order relation). Obviously, a choice is equivalent to itself (**Reflexivity of the order relation**). Finally, contradictory options are not allowed. It makes no sense to say that option 4 is preferable to option 5, that option 5 is preferable to option 6 and that option 6 is preferable to option 4 (**Transitivity of the order relation**).

B.2. Options Preferred to all the Options in a Specific set

One can focus his attention on a subset of options among the possible ones and try to decide which options are preferred to all the options in the given subset.

The reader may have difficulties in reading the previous sentence. Therefore, it is useful to replace it with the following one

One can focus his attention on a subset T of options among the possible ones and try to decide which options are preferred to all the options in T.

A nominalistic effort, i.e. calling T the considered subset of options, made the understanding of the sentence much simpler!

Definition We will call upper bound of a subset T included in S an element s of S such that for each t belonging to T, it is verified that $t \leq s$. The reader must remark that an upper bound of T may belong but also may not belong to T.

To continue with a more futile example, let us imagine that one has the following set of actresses:

A1. Anna Falchi, A2. Brigitte Bardot, A3. Claudia Cardinale, A4. Angelina Jolie, A5. Amber Heard, A6. Kate Moss, A7. Marylin Monroe.

Imagine then asking a decision-maker to express a rational choice about the attractiveness of the actresses just listed. For each pair of the actresses in the list, it will be necessary to say which one he prefers or if he considers that they have the same attractiveness. The attentive reader will want to hear specified that the comparison of the attractiveness is intended accomplished by assuming each element of the previous list in the maximum splendour of its beauty, and not considering the relative beauty in a given instant of time.

Even if this example may seem at first sight not relevant in social sciences, one can change his mind by simply considering that the Miss America or Miss Universe selection is based on the choice among the participant women, as made by the members of a jury. Now, it is clear that each jury member has, in his mind, a specific order of beauty. However, the order chosen by a jury member may differ from the order chosen by another jury member. How can the jury reach a decision? This is a problem, which belongs to the class of problems studied by Condorcet and Arrow. Everybody has heard long discussions among men and women concerning the ordering in the beauty of specific groups of human being: it seems that this ordering process has obsessed humanity since it first appeared on earth. Sexual choice, i.e. the choice of the father or mother of our children, is based on a similar ordering procedure.

A possible choice (which has been formulated by the author) is given by the following inequality string:

$$A6 < A5 = A4 < A2 = A3 < A1 < A7.$$

Note the economy of this notation.

Instead of listing one by one all the relationships of preference, by using the transitive property and the fact that the considered relation is a total order relation, the previous inequality string allows to deduce each of the relationships of preference individually. For example, $A6 < A7$. It should also be noted that, in the choice considered, A3 is equivalent to A2 and that A7 is preferred to all other options. If T is the set of options formed by the options A2, A3, A4, A6, then both A1 and A7 are upper bounds for T.

Definition (*Supremum*) Let $M(T)$ be the set of all the upper bounds of T. We will call the element m supremum of T in the set S if it is an upper bound of T, and if it is such that for each t belonging to $M(T)$, the inequality $m \leq t$ holds.

In short, a supremum is smaller than or equal to any upper bound.

Let T be the set of options A2, A3, A4, A6. Clearly, $M(T)$ is formed by the choices A1, A7, A2, A3. It is equally clear that both A2 and A3 are a supremum of T.

As a further example, considering the choice of President Kennedy, option 3 is a supremum of the set T formed by options 4, 5, and 6.

B.3. Rational Choice, in the Sense of Arrow

Arrow wants to limit his attention to a particular class of Arrow choices. He adds the adjective «rational» to the word «choice», to add a further particular specification to the choices he wants to consider.

Definition We will call **rational choice (according to Arrow)** of the individual **i** or of the whole group G a total order relation in S that verifies the following property:

For each subset T of S, there is at least one option in T which is its supremum.

Remark that such a supremum is not necessarily unique!

The assumption of rationality of economic or social agents or decision-makers is considered one of the main limits of modern economical and social sciences.

Many scientists have debated for a long time this assumption. Sceptical sociologists or economists, or in general, those who deny the human possibility to describe social phenomena with mathematical theories, claim that social agents are not rational, or may decide to behave in a non-rational way.

The debate, originated by these sceptics, attacks and mistrusts a series of hypotheses made by many of the most outstanding scientists in the last centuries, when dealing with the formulation of mathematical models for social phenomena. These assumptions are all very similar, in spirit and method, to the assumption implicit in the aforementioned definition by Arrow.

Arrows claims that a choice is rational if for every subset of possible options, there must be always a possible option, which is preferred to all of them and belongs to the selected subset. One should not believe that rationality is fully described by such an assumption: Arrow needed to chose a word and he made his choice. It is a pure nominalistic question, once precise definitions are given. Every scientist is aware of the fact that the concept of rational choice (following Arrow) can be replaced by another definition of *to-be-considered* choice, by a future mathematician who will be able to improve Arrow's analysis.

Those who criticise Arrow on the basis of a purely nominalistic issue are not the first and will not be the last.

Sextus Empiricus, i.e. "Sextus the Empiricist", a philosopher of the sceptical school who lived in the second century CE, is probably one of the first deniers of the predictive capacity of mathematical modelling. Sextus denied this capacity in every aspect of human knowledge: in physics, chemistry, medicine and all social and economical sciences. Nowadays, none dares to assume, openly, the extreme position held by Sextus. The modern sceptical school limits itself to deny the possibility to describe social phenomena by means of the scientific method. It has to be remarked that, unfortunately, also in medicine, the deniers of science are regaining more recently some positions: consider for instance the positions held by no-vax fanatics.

We do not want here to claim that the models developed up to now by applied mathematicians are sufficient to describe, as effectively as done for physical phenomena, also social and economical phenomena. We simply believe that we are in a transitory phase that is very similar to the one in which Galileo started to understand physical natural phenomena by means of geometry and mathematical theorems.

We believe that eventually a scientific understanding of social phenomena will be attained: of course, it is very difficult to imagine when this result will be actually attained.

We try to go a little bit deeper in aforementioned analogy: Galileo did start studying the motion of bodies in space by assuming that they can be modelled by a material point, i.e. a body without any finite extension. A sceptical scientist, follower of Sextus, may claim that Galileo's approach is useless, as in practice every body has finite dimensions. Moreover in his study of trajectories of projectiles, Galileo, and his followers, did assume that there is no perceptible resistance exerted by air to the projectiles' motion. Again the same criticism as before could be raised, to disqualify Galilean simplifications as completely illogical and Galilean theories completely useless: every body is always subject to frictional forces, when it moves!

Indeed, while Galilean analysis was not surely sufficiently careful to describe in a quantitative way the trajectory of a cannonball, it nevertheless represented an important conceptual step towards the full understanding of the mechanics of extended bodies subject to frictional forces. The preliminary and qualitative predictions of Galilean analysis paved the way towards the most advanced studies performed by, among the others, Newton, Hooke, Lagrange, Hamilton and Rayleigh. Without the first simplifying assumptions accepted by Galileo, it would have been impossible for his followers to get, by refinements of the necessarily imprecise first modelling attempts, the surprising prediction capacity which classical mechanics eventually has allowed us to master.

Therefore, we expect that, in future developments of the mathematically modelling of social phenomena, the simplifying assumption, which were temporarily put forward by Arrow, will be removed.

The role of these assumptions is to give us a first-level understanding of phenomena, so to have a guide for further theoretical developments.

Let us leave, therefore, the sceptical economist complains about irrationality of human choices; and let us investigate the surprising extension of our understanding obtained thanks to the Arrow's results. Their refinement will be the job of the new generations of scientists, who will surely improve Arrow's methods and definitions. It is also to those who may aspire to continue Arrow's works that this work is directed.

B.4. (Strictly) Preferred and Indifferent Options

Definition (*Preferred options*) We will denote with the symbol Pref(T) the subset of T whose generic element (a) is an upper bound of T and (b) belongs to T.

The Pref(T) set, which is included in T by definition, represents the set of all options in T that are preferred or indifferent to all the other options in T.

To go back to our last example: always for the set T constituted by A2, A3, A4, A6, we have cleared that the set Pref(T) is constituted by the set made by the two elements A2 and A3.

Observation Note that the Pref(T) set is not empty because of the definition of rational choice as given by Arrow. In addition, such a set can generally consist of more than one element: in fact, there may be several preferred options in T that are indifferent to each other.

Going back to the example dealing with the Cuban crisis: as is well known, President Kennedy made a choice that placed as his highest priority the option (3) among the options listed above. In other words: in the order relation chosen by Kennedy, the option (3) was preferred over all other options. So (3) was the only supremum in the set of all possible options.

The reader must recall once more that, in a rational choice, all the preferences between all possible pairs of options must be specified, even if the possibility that two options are considered indifferent (or equivalent) is contemplated.

Definition **We will say that s is indifferent to r**, and we will use the notation $s = r$ when we have at the same time that

$$s \leq r \text{ and } r \leq s.$$

In fact, when this last circumstance is verified, it is clear that none of the options s and r, with respect to the made choice, is preferred on the other one.

Definition Fixed a rational choice we can determine if between the two options r and s in S there is one, which is really preferred: in fact, we will say that r is **strictly preferred** to s (and we will use the notation rPs) when it is not true that $r \leq s$.

Therefore, if a rational choice is made, then it is possible to introduce in S two further relations which we have denoted respectively with the symbol P and with the symbol =.

These relations indicate preference and indifference in S as established by the considered Arrow choice and are called strict preference relation and indifference relation, respectively.

The reader will easily realise that a relationship of strict preference is not a total order relationship: in fact, a relationship of preference is neither reflexive (no choice is strictly preferred to itself) nor does it verify the condition of totality. In fact, it is possible that two options are indifferent and therefore neither the first is preferred to the second nor the second is preferred to the first: clearly, it is not possible to establish a relation of strict preference between two indifferent choices!

One can be easily persuaded that the following simple propositions are true.

Proposition 1 Given a rational choice $\leq$, if s and r are two options in S, then either $r \leq s$ or rPs.

In words of natural language, i.e. without using the formalism of mathematical notation, this statement can be rephrased as follows:

Given two options among the admissible ones, and a rational choice following Arrow, then two possibilities only are given: (i) the first option is preferred to the second one, or (ii) the second option is preferred or equivalent to the first one.

Proposition 2 Let s, r, p be three options in S. If $r \leq s$ and rPp, then sPp.

We can rephrase also this second proposition:

Given three options and a rational choice then if the option r is preferred to the option p and s is preferred or equivalent to r, then the option s must be preferred to the option p.

B.5. Votes in a Social Group

We want to consider the determination of the choice in a group as a decision that considers, as a starting point, the individual choices of its members. Therefore, we will need to address the problem of combining many choices to get the social choice.

As a consequence, we need to develop a symbology capable to distinguish between two (or more) different choices.

This distinction was not necessary up to now.

We will therefore consider, in what follows, two different choices: (≤ 1) and (≤ 2). In other words, we call (≤ 1) the first considered choice and (≤ 2) the second considered choice. Again it is important, to proceed in our reasoning to choose a good notation!

Definition Let us consider two rational choices (≤ 1) and (≤ 2) and one option p in S. We will say that

(≤ 2) **strengthens the preferences for p expressed by** (≤ 1)

(leaving the other preferences unchanged), and we will use the notation

$$(\leq 1)p(\leq 2)$$

if the following conditions are verified:

- For each pair (r, s) of options taken in S, both different from p, it results that

$$r(\leq 1)s \text{ if and only if } r(\leq 2)s.$$

- If s is an option in S, it results that

$$s(\leq 1)p \text{ implies that } s(\leq 2)p.$$

- If s is an option in S, it results that

$$pP(\leq 1)s \text{ implies that } pP(\leq 2)s.$$

where with the symbols $P(\leq 1)$ and $P(\leq 2)$, we mean the preference relation induced, respectively, by the choices (≤ 1) and (≤ 2).

The reader will easily realise that the terminology introduced is intuitively well-founded: in fact, the first bullet condition states formally that, for all pairs of options in S different from the option p, the ordering expressed by (≤ 1) coincides with the ordering expressed by (≤ 2). Furthermore, the second and third bullet conditions can be interpreted as a whole by saying that if the option p was preferred over a given option s in the choice (≤ 1), then it will still be preferred over s in the choice (≤ 2), and that if option p was indifferent over s in the choice (≤ 1), then it will be strictly preferred or indifferent to s in the choice (≤ 2).

The idea behind the previous definition is clear. *One has two choices. These two choices leave unaltered the order relation of all pairs of options except those involving a precise option, which we have called p. Instead, the order relation involving p increases, with the second choice, the preference for p as expressed by the first choice.*

Notation. We will denote with the symbol $\leq$i a rational choice of the individual i and with the symbol $\leq$G a rational choice of the group G.

As the reader now became accustomed to the notation, we will now skip the parentheses for denoting the choices when this is not causing ambiguities.

A vote is the set of choices expressed by all the N individuals in group G. More formally:

Definition (*Votes*) Let us consider the set V of all the N-tuples of rational choices that can be expressed by the individuals of the group G. An element of such a set is formed by the list (≤ 1, ≤ 2, …, $\leq N$) formed by the choice ≤ 1 of the individual 1, by the choice ≤ 2 of the individual 2, … and by the choice $\leq N$ of the individual N. Each individual has a total freedom in expressing his rational choice. Therefore, a generic element of V is a list of as many rational choices as the number of individuals in G. **Let v be an element of V: we can call v a vote of the individuals in G. V represents the set of possible votes.**

The attentive reader will remark that we continue to introduce a notation for making our phrasing easier. When we write $\leq i$ and we say that $i = 1, 2, …, N$ (in words that the variable i can take the values 1 or 2 or 3 or … or N), we mean that we are considering the set constituted by exactly N choices in the considered sequence. We consider an ordered set of choices: the first, the second, the third and so on. We mean that the fact that a choice is at the first place or at the fourth place, for instance, is relevant in our reasoning. This ordered set of choices has been indicated by the symbol (≤ 1, ≤ 2, …, $\leq N$).

Definition Let us consider two votes v and v': generalising the definition given for rational choices, we will say that the vote

$$\mathbf{v'} = (\leq'\mathbf{1}, \leq'\mathbf{2}, …, \leq'\mathbf{N})$$

strengthens the preferences for p expressed by

$$\mathbf{v} = (\leq\mathbf{1}, \leq\mathbf{2}, …, \leq\mathbf{N}),$$

if for each $i = 1, 2, …, N$, we have that ($\leq i$) p ($\leq' i$)

In other words, the vote v' strengthens the preferences for option p over the vote v when each individual i in the vote v' expresses a choice which strengthens his preferences for option p as expressed in the vote v.

Definition We will say that **two votes v′ and v order in an equivalent way the options in the subset T of S** when

for each pair r and s of options in T and for each $i = 1, ..., N$, we have that $r(\leq i)s$ if and only if $r(\leq'i)s$.

In other words, two votes order the options in T in an equivalent manner when all the individuals of the electoral body in the two votes express the same strict preferences and indifferences when considering the options listed in T.

B.6. Social Choice Function and Its Properties

We can now define the concept of social choice function.

Definition We will call (following Arrow) **function of social choice** a function that has as domain the set V of all the possible votes of the group of individuals G on the options represented by the set S and as codomain the set D of the possible rational choices of the group G. A function of social choice maps every vote into a social choice.

A possible rigorous formulation of the problem of Condorcet is now possible:

What are the functions of social choice that best represent democratically the will of the electoral body G?

It is much more precise to say that we have just given, among the many other possible ones, a specific rigorous formulation of the problem of Condorcet: the formulation conceived by Arrow! As Condorcet conjecture was formulated necessarily in an informal way, it is conceivable (and also desirable, as we have commented in the section dedicated to Montesquieu) that another specific formulation may be formulated.

The reader, who did manage to follow our definitions up to now, became quite familiar with the mathematical method. Therefore, he understands that the question we have asked requires, in order to be well posed, some more definitions: those which allow us to specify what the expression "represent democratically" precisely means.

Notation Let F be a function of social choice. Remember that, for the given definitions, if v is a vote, then $F(v)$ is the rational choice of the social group G, i.e. the decision of G, as ruled by F. With the symbol $F(v)$, we denote the choice obtained by applying the rule F to the set of choices expressed by the vote v.

The first assumption accepted by Arrow for a function of social choice is the most technical one, and also the one that can be questioned the more by the most sceptic readers.

1. Condition of Independence of Irrelevant Alternatives

Let v' and v two votes of the group G. Let $\text{Pref}(T, F(v))$ and $\text{Pref}(T, F(v'))$ be the options, in the subset T (included in S), which are strictly preferred or indifferent to all the other options in T when considering, respectively, the decisions $F(v)$ and $F(v')$.

We will say that **the social choice function F is independent of irrelevant alternatives** when

For each T included in S,
if v and v' order in an equivalent way the options in the subset T

$$\text{then } \text{Pref}(T, F(v)) = \text{Pref}(T, F(v')).$$

This assumption can be interpreted as follows. *Let us imagine that two different votes change their preferences for an option outside the set of options in T. It is assumed that the rule F, when applied to these two votes, cannot change the preferences in T if the preferences of an option outside T are changed.*

Roughly speaking, the rule F orders the options inside T in a way, which is not affected by the preferences of the options outside T.

This assumption has been, for a long time, debated and criticised and shows, clearly, some limits when the results by Arrow must be applied in practical cases.

For instance, imagine that the social choice has to express a preference among six candidates, A1, A2, A3, B1, B2 and B3. Assume that the A candidates belong to the same party, and the same is true for B candidates. The rule often used in forming the presidency committees in a Senate (the highest Chamber) is that the candidate who gets more votes becomes President. Assume that the candidates for this presidency are A1 and B1. The social choice rule for the Senate, may, however, be conceived to balance the power given to a certain party by imposing that the presidency of the committee for the control of secret services cannot be given to a member of the same party to which the President of the Senate belongs. Therefore in the ballot between A2, A3, B2 and B3, the social choice change its output depending on the results of the vote choosing between A1 and B1. Such a rule is clearly NOT independent of irrelevant alternatives. Of course, a solution to this difficulty is simple: one must split the vote into at least two steps: in a first step, the Senate President is elected, and in a subsequent step, it is the President of the secret services committee to be chosen.

There is also a second condition that Arrow imposes to a function of social choice.

2. Monotony Condition in Preferences

Let v' be a vote that strengthens the preferences for some option p as expressed by the vote v.

We will say that the **function F enjoys the property of faithfully representing the strengthening of preferences** (or more briefly that it **is monotonous in preferences**) when

for each r in S if $pP(F(v))r$ then $pP(F(v'))r$.

We do not believe that this last assumption may be criticised by any sceptical philosopher. Such an assumption seems essential for any democratic function of social choice.

Of course, a single individual should not be able to determine the output of a vote. Therefore, the following condition cannot be accepted for a democratic social choice.

3. Condition of existence of a dictator

We will say that **a function of social choice F is dictatorial** if

there is an individual C in G such that for each vote v in V and for each s and r

$$sP(\leq C)r \quad \text{implies} \quad sP(F(v))r,$$

where $\leq C$ represents the choice of the individual C.

A true democratic function of social choice must treat all voting individuals equally. Therefore, it should verify the following condition.

4. Condition of equivalence of the voters (or condition of democracy) of the function of social choice.

Given two individuals i and j in G and a vote $v(i, j)$ in which $\leq a$ and $\leq b$ are the choices of i and j, respectively. We will call a vote $v(j, i)$ a vote in which the individual i chooses $\leq b$ and j chooses $\leq a$. In other words, in $v(j, i)$ the choices made by i and j in the vote $v(i, j)$ are exchanged.

We will say that **a social choice function F considers voters equally** if
For each pair (i, j) of individuals in G and for each s and r in S

$$sP(F(v(i,j)))r \quad \text{if and only if } sP(F(v(j,i)))r.$$

Clearly, the existence of a dictator in a function of choice means that this function of choice is not democratic, in the sense of the previous definition.

The attentive reader will recall that we have already discussed the concept of imposed choice. Very often, in many electoral systems a specific criterion of choice, in the event of a tie, is established: the oldest of the candidates having obtained the maximum of votes is elected. Also, in the bonobo societies the hierarchy among the females, from which the hierarchy among their male children derives, is established by an imposed choice: that of seniority. Note that the bonobos are, by far, the most peaceful of the non-human apes. We conjecture that this is possibly true because the existence of such an imposed hierarchy, which avoids any kind of physical confrontation for establishing the group internal pecking order. We shall also notice that the bonobo is, always among apes, those at greatest risk of extinction: possibly, this circumstance may be related to their incapacity to confront conflicts.

Nature, and human social traditions, teaches us that there are some social choice functions in which some preferences are imposed a priori. It is therefore not surprising that Arrow introduces a last definition.

5. Condition of Imposed Preferences

We will say that **a social choice function F is imposed or equivalently that it has some imposed preferences** when

there is at least one pair of options s and r in S such that, for every vote v, $sF(v)r$.

The examples of imposed choices in human societies are very numerous. It is very suggestive that Western civilisation, which has been following many imposed preferences in social choices did manage to understand and formulate explicitly the concept only when Captain Cook observed the «primitive» Tongan society. He introduced in all modern languages the word **taboo**, which perfectly characterises the set of options, which are forbidden by an imposed social choice.

The fact that taboos depend on the social group considered at the time of observation and are not universal was not remarked explicitly until very recent times. Only very recently, for instance, it became possible for a divorced man to try to be candidate for the highest political offices. The taboo that forbids electing a man (or a woman) who was not capable to manage to remain married has been finally removed.

B.7. Arrow's Theorem on Possible Functions of Social Choice

We are now in the most suitable position to enunciate Arrow's theorem on possible functions of social choice:

Consider a set of possible options S consisting of at least three options. Let F be a social choice function that associates a decision (i.e. a social choice) to each vote. If F verifies the conditions of independence from irrelevant alternatives and monotony in the preferences then either F has imposed choices or it is dictatorial.

For a demonstration of Arrow's theorem, we refer the reader, for example, to Arrow's work and the Fishburn manual cited in the bibliography.

We hope that the reader when arriving at this final chapter increasingly appreciated the value of Arrow's work in the reorganisation of Condorcet concepts and in the needed mathematical formalisation.

While surely Arrow has proposed a precise formulation of the concept of rational choice, preference, individual choice and social choice, we do not claim that Arrow's remarkable efforts have terminated the problem of finding a democratic organisation of human societies.

Arrow's result is a result of impossibility: if there are at least three options among which the individuals of the group G must choose, there is no function of social choice which faithfully reflects the expression of the vote (i.e. following the Arrow's definitions, a function which is monotonous in preferences and which is independent of the irrelevant alternatives) and which is, at the same time, neither dictatorial nor with imposed preferences. The Condorcet search of an efficient and truly democratic system is not yet ended. As discussed in a previous chapter, we hope that the ideas of Montesquieu can allow us to progress in this search.

Annotated Bibliography

Acharya, S., & Shukla, S. (2012). Mirror neurons: Enigma of the metaphysical modular brain. *Journal of Natural Science, Biology and Medicine, 3*(2), 118–124.
*This paper is really clear and drives the readers in some recently unveiled mysteries of human (and other apes') brains. If one has some education in computer science he will discover that, actually, primates' brains seem to be analog computers. The action of understanding other individuals' behaviour is attained by building some **specific hardware** in the brain, as a consequence of the learning process.*
Acton, H. (1997). The last bourbons of Naples (1825–1861). Firenze: Giunti. ISBN 88-09-21256-8.
This historical work is of great importance, we believe. Indeed the education and origins of the author are mainly British, and he had no sentimental reason for giving a distorted version of the events which led to the end of the Kingdom of Two Sicilies.
Arnol'd, V. I. (1998). On teaching mathematics. *Russian Mathematical Surveys, 53*(1), 229.
This masterpiece in the philosophy of mathematical physics is a major source of many deep and clever considerations about science and its perspectives. Not all the sharp conclusions by Arnol'd has to be a priori accepted, however, without exerting a wise critical analysis.
Arrow, K. J. (1950). A difficulty in the concept of social welfare. *Journal of Political Economy, 58* (4), 328–346.
This is the masterpiece of applied mathematical science that inspired the greatest part of the present essay. It is innovative but also relatively simple: everybody having a basic education in Set Theory can read it. It proves that great results do not need to be hidden by the fogs raised by a complex jargon.
Aureli, F., & de Waal F. (Eds.). (2000). A difficulty in the concept of social welfare natural conflict resolution. University of California Press.
This is another masterpiece. Many ethologists sustained the opinion that human nature is only violent and inclined to resolve conflicts based on the so-called «law of the jungle». Instead experimental observation of primates behaviour proves that, in nature, the search of compromise and the behaviour which looks for the solution of conflicts without the use of force are frequent and successfully applied, not only among bonobos, but also among chimpanzees and gorillas. Therefore one cannot claim that the «natural state» of humans is a permanent fight for survival.
Calà Ulloa, P. (1876). Di Carlo Filangieri nella storia de' nostri tempi. Napoli: Tornese.
This work is written by one of the most serious, honest and educated magistrates of the Kingdom of Two Sicilies. In this work the shining figure of Carlo Filangieri is delineated with all its shadows and lights.

© Springer Nature Singapore Pte Ltd. 2019
F. dell'Isola, *Big-(Wo)men, Tyrants, Chiefs, Dictators, Emperors and Presidents,*
https://doi.org/10.1007/978-981-13-9479-9

Ciccarelli, C., & Fenoaltea, S. (2012). Through the magnifying glass: Provincial aspects of industrial growth in post-unification Italy. *Economic History Review 66*(1), 57–85.

Daniele, V., & Malanima, P. (2007). Il prodotto delle regioni e il divario Nord-Sud in Italia (1861–2004). *Rivista di Politica Economica* 267–315.

The previous two works are a clear and irrefutable proof of the fact that the Kingdom of Two Sicilies was a wealthy and rather modern State, with a strong economy. The data (and the economical reasoning) presented there have never been confuted by any serious scientist. They support in an authoritative way the conclusions presented in the present essay.

Dardanoni, V. (2001). A pedagogical proof of Arrow's impossibility theorem. *Social Choice and Welfare, 18*(1), 107–112.

In this work a simple, but rigorous, proof of Arrow's Theorem is presented. To our knowledge, this is the most suitable reading for those who intend to delve more in the mathematical details of such proof.

De Cesare, R. (1969). La fine di un regno, prima edizione 1895, Longanesi & C, Milano.

The young De Cesare was a personal witness of the event's which he describes. He had a great culture, he was an expert journalist: Deputy in the first Italian Parliament, he was then nominated Senator. His monumental work is a careful list of sources and historical evidence. Reading this work may be boring, and many historians claimed that its scientific value is not really relevant. Instead we believe that it is an invaluable source of «facts» without «opinions». De Cesare gave us the needed historical evidence for supporting the theoretical interpretation of the facts, which led to the unification of Italy. These facts must be taken into account by new generations, while facing the process of unification of Europe.

de Waal, F. (1989). Food sharing and reciprocal obligations among chimpanzees. *Journal of Human Evolution, 18*(5), 433–459.

de Waal, F. (1990). Peacemaking among primates. Harvard University Press.

de Waal, F. (1995). Bonobo sex and society. *Scientific American, 272*(3), 82–88.

de Waal, F. (1996). Good natured (No. 87). Harvard University Press.

de Waal, F. (2006). Our inner ape: A leading primatologist explains why we are who we are. Penguin.

de Waal, F. (2007). Chimpanzee politics: Power and sex among apes. JHU Press.

de Waal, F. (2009). Primates and philosophers: How morality evolved. Princeton University Press.

de Waal, F. (2010). The age of empathy: Nature's lessons for a kinder society. Broadway Books.

de Waal, F., & Lanting, F. (1997). Bonobo: The forgotten ape. University of California Press.

The previous (incomplete) list of works by de Waal represents a major contribution to the self-awareness of the nature of human beings. They are so carefully, precisely and rigorously written that we believe it is rather difficult to deny de Waal's vision about human nature. Starting from observations about our closest relatives, i.e. Bonobos, Chimpanzees and Gorillas, the analysis presented by de Waal gives a comprehensive understanding of human ethology. One has to remark that the conclusions presented by de Waal are not as negative as those initially proposed by those ethologists who first observed the violent social behaviour of chimpanzees. De Waal is not denying the most violent aspects of our nature, but he proves that they are compensated by other aspects, which are surprisingly altruistic and full of empathy. He talks about «bipolar» apes, i.e. humans, showing extreme behaviours: humans appear sometimes extremely «good-natured» and sometimes extremely «bad-natured». Probably the considerations, which we have presented about Nash mixed strategies, may be of help in understanding the described bipolarity.

Filangieri Fieschi Ravaschieri, T. (1902). Il generale Carlo Filangieri, principe di Satriano e duca di Taormina, 53. Milano: Treves.

The daughter of Carlo Filangieri, being afraid that her father's deeds may be forgotten, gathered all the documents which she could find and published this precious biography. Of course many northerner historians sustained that the work of the daughter could not be considered reliable when describing the life of the father. Luckily for historical truth we also have the works by De Cesare and Calà Ulloa.

Fishburn, P. C. (1973). The theory of social choice. Princeton University Press.
This textbook is the most comprehensive that we have found about Arrow's theorem. Everybody who needs to understand fully its technicalities should peruse it.
Miller. (2000). The mating mind: How sexual choice shaped the evolution of human nature. Heinemann. ISBN 0-434-00741-2.
This book is enlightening in many aspects. However it is also misleading in others. We believe, however, that reading it may be really useful, if one wants to understand human nature. When the author refuses to try to apply rationality to describe human behaviour (as for instance homosexuality) the reader will see the power of cultural taboos, which even serious scientists, sometimes, do not manage to infringe.
Mitani, J. (1995). Kanzi: The ape at the brink of the human mind. *Scientific American, 272*(6). ISSN 0036-8733 (https://www.worldcat.org/issn/0036-8733).
In this work a scientific and rigorous presentation of the intellectual performances of Kanzi is made available to non-specialistic audience. Reading this book is extremely useful for us: it is like discovering the existence of another intelligence, very similar, but different from ours. It is really useful to compare the behaviour of Kanzi with the behaviour of a 2/3 years old child or with the behaviour of a Big-Man. Linguistically speaking Kanzi did not manage to overcome a 4 years old child: however many politicians may learn by him how a leader should behave.
Nitti, F. S. (1903–1958). Scritti sulla questione meridionale (Vol. 1). Laterza.
Nitti has been an influential statesman, a famous economist and a man who never denied or hid his Southern origin. His scientific analysis of the economical situation in Italy before and after unification is another cornerstone in the fair reassessment, which is needed, in this subject. Nitti had access to all data (he has been Italian Prime Minister), he had the required scientific standing (he has been professor of Science of Public Finance at the Università di Napoli) and he could try to act to improve the situation with the political action of his government. He has been one of the most eminent representatives of Meridionalism (Italian: Meridionalismo). With the word Meridionalism is indicated the scholarly study of the political, economical and social situation of Southern Italy, as a result of its inclusion in Italian State. This study has, as its main target, the search of solutions of the problems caused by this forced inclusion to try to bridge the economical, social and cultural gap established, since then, between North Italy and South Italy. The scholars and politicians who worked in this field are referred to as "Meridionalists" (Italian: Meridionalisti).
Patterson, F., & Linden, E. (1981). The education of Koko. New York: Holt, Rinehart and Winston.
Koko has been a Gorilla (she died in June 2018) and scored an intelligence quotient between 70 and 90, when using different scales. She, following many experts, has proven to be able to use effectively around 1000 words, in a specifically designed sign language, and understand nearly 2000 words uttered in English. Following Mary Lee Jensvold, an expert in primates ethology: "Koko ... [used] language the same way people do". This book tells the story of her education. Everybody who still doubts about the capacity of non-human primates should read it to be persuaded on the contrary.
Perrone, N. (2009). L'inventore del trasformismo. Liborio Romano, strumento di Cavour per la conquista di Napoli. Soveria Mannelli (Catanzaro): Rubbettino. ISBN 978-88-498-2496-4
In this book many sources about the life and the actions of Don Liborio are gathered and reorganised. This is a precious tool for clarifying Don Liborio true role in the process of Italian unification.
Pinker, S. (2008). The sexual paradox: Extreme men, gifted women and the real gender gap. Scribner.
Are women and men different? How they manage power? Which is their nature? These questions are beautifully answered in this enlightening book. Very important for future social engineering!
Preston, S. D., & de Waal F. (2002). Empathy: Its ultimate and proximate bases. *Behaviorial and Brain Sciences, 25*(1), 1–20.

This technical paper shows that empathy is NOT a cultural structure which higher education is imposing to some humans. In fact it is an ethological feature which humans share with other primates. It has an evolutionary importance and meaning and belongs to our innate psychological structure.

Russon, A. E., Bard, K. A., & Parker, S. T. (Eds.). (1998). Reaching into thought: The minds of the great apes. Cambridge University Press.

This collection of essays is one of the most persuasive contributions to the thesis of the existence of a kind of abstract thought in great apes mind. Facts, indeed, show that human intelligence is not appeared suddenly only in our species. A gradual process slowly occurred during evolution and eventually led to our intelligence: some sparks of our ancestors' capacities can be observed also in our genetically closest relatives. Actually in the mind of great apes all those features of intelligence, which found their expression in our minds, can be found «in embryo».

Sahlins. M. (1963). Poor man, rich man, big man, chief; Political types in Melanesia and Polynesia. *Comparative Studies in Society and History, 5*(3), 285–303.

This book is a careful description of politics in human groups before the invention of the concept of «state». It is instructive to see that in these human groups the social dynamics is very similar to the one observed in groups of apes and in the groups of university professors.

Savage-Rumbaugh, S., & Lewin, R., (1994). Kanzi: The ape at the brink of the human mind. Wiley. ISBN 0-471-58591-2.

Savage-Rumbaugh, S., Taylor, T. J., Savage-Rumbaugh, E. S., & Shanker, S. (1998). Apes, language, and the human mind. Oxford University Press on Demand.

The previous two books describe the evidence gathered by the group of Susan Savage-Rumbaugh about the manifestation of linguistic capacities in apes and in particular about the capacities of Kanzi. The conclusions seem very well-grounded and have changed our understanding of our own mind.

Savarese, G. (1862). Le finanze napoletane e le finanze piemontesi dal 1848 al 1860 (Vol. 2875). tip. di Gaetano Cardamone, Google books.

In this book it is discussed the situation of the finances of the Kingdom of Two Sicilies before the unification of Italy. A comparison with the finances of the Piedmontese Kingdom is performed and, based on what seem to be a solid argument, it is concluded that before unification South Italy was not in a bad economical situation, and that probably it was richer than North Italy.

Service, E. R. (1975). Origins of the state and civilization: The process of cultural evolution. Norton.

In this book the process, which transformed human societies into states, is carefully described and the true differences between human and apes social groups are highlighted.

Viesti, G. (2013). Il Sud vive sulle spalle dell'Italia che produce. Falso! Laterza Bari-Roma.

In this essay-pamphlet it is documented how northerner Italian economical forces managed to exploit the effort started by Italian State to help the economical emancipation of South Italy. Actually the greatest part of the investments intended to help the growth of South Italy were (illegally) diverted to the North, obtaining the paradoxical effect of increasing the North–South economical gap.

Viola, P. (1994). È legale perché lo voglio io Attualità della Rivoluzione Francese. Laterza.

This book is astonishing for a modern reader who has superficially studied the European history. Absolute Monarchs were not free to do what they wanted without any limit or control. A not-written constitution was ruling their behaviour as «they were not despots» as the sultans and other kings outside Europe. The noblesse, the clergy and also, in a certain measure, The Third State had many ways to limit the monarch power and discretion.

Wade, N. (2006). Before the dawn. Recovering the lost history of our ancestors. Penguin

In this book the natural history of human being is described, based on the most recent knowledges in biology, evolutionary ecology and genetics. Reading it, one gains a persuasive understanding of the reasons for which human nature is what it is.

Printed in the United States
By Bookmasters